MACRO STRUCTURAL PERSPECTIVES IN INDIAN AGRICULTURE

Macro Structural Perspectives in Indian Agriculture

MANABENDU CHATTOPADHYAY

GIAN PUBLISHING HOUSE
Delhi-110 007

GIAN PUBLISHING HOUSE
29/6, Shakti Nagar
Delhi-110007

ISBN 81-212-0241-8

Printed in India
at
Amar Printing Agency.
Delhi.

Published by Mrs. Gayatri Garg for
Gian Publishing House, Delhi and

To
My Elder brother
SHRI SHYAMALENDU CHATTOPADHYAY
from whom
I received encouragement and blessings in my academic endeavours

Preface

In a predominantly agrarian economy like India's agriculture plays an important role in any discussion of the nature of thc Indian economy. There is a large volume of literature on Indian agriculture in which different aspects of agrarian conditions and relations have been discussed. Yet most of such discussion is characterised by pre-conceived assertions than by empirically informed analysis, and further, they fail to analyse the structural similarities and differences of different regions within a comparative perspective.

The theoretical orientation of this study seeks to understand and comprehend the macro structure of agriculture for some regions of India in terms of a few major parameters (viz. productivity, cropping pattern, farm profit, tenancy, labour etc).

Many of the findings presented here have been reflected in different articles published in newspapers, weeklies and journals. It has been felt that a consolidation of all these important findings on structural aspects of agriculture might prove helpful for research workers who might be interested in carrying out similar studies in other regions of India.

I have divided the study into two parts. In part I, I have presented some empirical evidence on some selected topics of interest for some regions of India, whereas in Part II, critical analyses of certain quantitative relations pertaining only to

West Bengal agriculture have been provided. In both parts, however, I have not refrained from attempting certain fresh formulations on quantitative relations of production which have not yet received proper attention by our theoretical economists. This is important for a proper understanding of the peculiarities of our agrarian economy. These peculiarities have been summed up in the Concluding chapter i.e. chapter (8) of this study and the structural constraints of rural development have been discussed in this perspective.

In the course of this work, I have benefited greatly from discussions with Professor Donald Von Eschen of McGill University, Danada, Professor Partha N. Mukherji of the Indian Statistical Institute and many other scholars of the ISI and the Calcutta University working in this field. I express my heart felt gratitude to all of them.

I take this opportunity to extend my sincere thanks to the authorities of the Indian Stat istical Institute for having facilitated my work in every manne r possible.

MANABENDU CHATTOPADHYAY

Contents

Part I

Indian Agriculture : Some Aspects

1

Growth and Stagnation of Agricultural Labour

1. Work Participation

Certain difference in concept and in definition have created difficulties in comparing the census data of various rounds. The census of 1961 used the 'usual status' definition of a worker which provided an inventory of persons engaged in any economic activity on a more or less regular basis. The reference period, excepting trade, business, professions or services, for which it was specifically a fortnight prior to the date of enumeration was broadly defined as 'the greater part of the working season' during which any one gainfully employed for at least one hour a day was considered to be worker. The 1971 census made a major departure by using a 'labour time disposition' criterion over the reference period (one year for those engaged in cultivation/livestock raising/household industries and plantations), and by defining workers as only those who spent the major part of their time in economic activities. Thus all irregular/marginal/part-time workers, by the time disposition criterion, came to be enumerated as non-workers.

At the 1981 census, there was an attempt to collect information on main workers and marginal workers separately. The 1981 census also used 'time disposition criterion' over the reference period (one year) and defined main workers as those who worked for a major part of the year preceding the enumeration.

Although the concepts of 'workers' over census rounds are not comparable, a measure of total working force in 1961 comparable with that of the 1971 and 1981 censuses is possible by including the workers who are engaged in the marginal or part-time activities in the 1971 or 1981 concept.

TABLE—1

Work Participation Rates 1981-1961 : All India

Type of worker with census year	*Rural*			*Urban*		
	Male	Female	Total	Male	Female	Total
(1)	(2)	(3)	(4)	(5)	(6)	(7)
Workers/Main workers						
1961	—	—	—	—	—	—
1971	53.62	13.44	34.04	48.82	6.68	23.34
1981	52.21	16.49	34.77	48.18	7.57	29.17
Workers/Main workers plus Secondary workers/ *Marginal* workers						
1961	58.30	31.42	45.09	52.37	11.16	33.47
1971	53.78	15.92	35.33	48.88	7.18	29.61
1981	54.32	23.89	39.46	49.70	10.64	31.41

Source : *Census of India 1981*, Series-1, India, Paper-3 of 1981, Provisional Population Totals.

Note : The term "work participation" used in this analysis refers to the proportion of number of workers to total population.

Table 1 has been prepared on the basis of the adjustments described above which shows the work participation rates during the Period 1961-81. Considering the data of main workers for the years 1971 and 1981, it is seen that work participation rates in both rural and urban areas have remained more or less constant during the period 1971-81. In the case of females, however, there is an increasing tendency in the work participation rate between 1971 and 1981 both in the rural and urban areas.

A comparison of work participation rates, the combined rates of 1981 with the work participation rates of 1961 (presented in the same table), would show that the 1961 rates are higher than the 1981 rates. However, if one were to compare the numbers rather than the rates as such, it will be noticed that the number of main workers and marginal workers together in 1981 has increased considerably over the number of workers in 1961 (i.e., the number of such workers has increased from 182.5 million in 1961 to 247.2 million in 1981).

Regarding the participation rates during the period 1961-71, some scholars (e.g.. Visaria 1976; Bardhan 1977 etc.) argued that whichever concept and set of adjustment factors are used, a comparison between 1961 and 1971 indicates a genuine decline in work participation rates—a decline that has been more drastic in the rural than in the urban areas, and among rural women in particular.

A demographic explanation for the decline in the overall participation rate during the period 1961-71 derives from an increase in the proportion of children over the decades (Visaria 1976). It has also been noted that rise in incomes and in the incidence of school enrolment that took place in the rural areas over the decade might have contributed to the decline in the worker participation rate for all ages conbined.

Regarding the decrease in female labour participation in the rural areas during the period 1961-71, Kalpana Bardhan (1977) put forward the view that many of the productive work that the women of farm families do (like winnowing, preparation of dairy products, storage of seeds and produce etc.,) are quite often not qualitatively different from some of the purely domestic chores and tend to be under reported. This sort of under reporting of female self-employment would naturally affect female participation rates. A very different argument is that as a consequence of technological changes taking place in agriculture female labour has been increasingly pushed out of farm labour into domestic activities (Billings and Singh 1969).

TABLE 2

Female work Participation rate : Statewise—1971-81

	1981	**1971**
Andhra Pradesh	27.87	24.16
Maharashtra	24.39	19.70
Madhya Pradesh	22.63	18.65
Tamil Nadu	22.57	15.09
Karnataka	19.23	14.20
Himachal Pradesh	18.80	20.79
Kerala	12.79	13.49
Gujarat	11.85	10.26
Orissa	10.88	6.81
Rajasthan	9.44	8.34
Bihar	9.16	8.88
Tripura	9.08	4.83
Uttar Pradesh	6.02	6.71
West Bengal	5.97	4.43
Haryana	4.82	2.41
Punjab	3.09	1.18

Source : *Ibid*, p. 8

Recent trend of increased work participation in general and female work participation in particular does not corroborate the views put forward by the scholars described above. The very different pattern of movement of the work participation rates in 1981 indicates that the declining trend in the work participation rates during 1961-71 is likely to have been more apparent than real. One possible cause for this may be that the nuances of the definition of earlier census rounds were lost while the data were collected. Doubts are stengthened by the fact that an increasing rate of female work participation is noticed during 1971-81 in the majority of the states in India (Table 2).

Furthermore, the data on per cent increase of main workers in 1981 over workers in 1971 in different states of India do not indicate any sign of declining trend of the proportion of workers either in rural or in urban India (Table 3).

TABLE 3

Per cent Increase of Main Workers in 1981 over Workers in 1971 by Some States

State	Rural			Urban		
	Male	Female	Total	Male	Female	Total
(1)	(2)	(3)	(4)	(5)	(6)	(7)
Andhra Pradesh	13.67	40.80	22.04	44.54	56.26	46.50
Bihar	14.53	26.45	16.27	39.53	45.21	39.94
Gujarat	21.90	47.10	26.40	44.90	50.03	45.38
Haryana	20.52	169.54	26.51	73.69	114.46	75.83
Himachal Pradesh	10.85	10.85	10.85	28.38	89.43	33.77
Karnataka	17.99	70.93	29.29	47.66	79.57	51.93
Kerala	3.32	11.47	5.28	25.66	26.87	29.47
Madhya Pradesh	19.01	49.90	27.17	56.60	81.92	59.64
Maharashtra	18.66	54.11	29.77	35.24	58.77	38.01
Orissa	13.95	89.99	22.26	62.82	99.33	66.42
Punjab	12.18	316.79	15.69	45.96	147.24	50.39
Rajasthan	19.40	47.78	23.29	57.71	89.16	59.69
Tamil Nadu	11.38	79.21	26.95	27.74	57.28	32.09
Tripura	28.82	152.34	38.95	44.97	112.20	52.25
Uttar Pradesh	12.96	8.12	12.44	56.55	89.02	58.19
West Bengal	21.52	65.08	25.08	23.18	78.05	26.25
All India	15.60	46.33	21.51	42.49	67.55	45.13

Source : *Ibid*, PP. 37-41

2. Growth of Agricultural Labourers

The distribution of main workers by the broad categories of 'cultivators', 'agricultural labourers' etc., for the country as a whole is provided by each census-round.

The following observations can be drawn from such data as presented in Table 4.

(a) An absolute rise of agricultural labourers from 31.3 million in 1961 to 55.4 million in 1981 has taken place during the recent decades. Proportion of agricultural labourers to total main workers reveal a tendency to increase definitely during the period 1961-71, but no particular trend of increasing in terms of proportion is noticed during the current census decade.

(b) The figures also show an absolute rise of cultivators during the period 1971-81. But the reverse is true during the period 1961-71. In terms of proportion, however, there shows a decreasing trend or cultivators over the entire time period.

(c) Although the number of both agricultural labourers as well as cultivators has been increasing during the period 1971-81, the intensity of increase in case of cultivators is more substantial than the agricultural labourers (as against 8 million increase of agricultural labourers, the cultivators has increased to 16 million). Thus, the rate of increase af cultivator category is about two times compared to the rate of increase of agricultural labourers during the period 1971-81.

While there are large increases both in the cultivator and labour categories in the number of workers over the period 1971-81, it can be seen from the same table that the number and proportion of cultivators decreased and the number and proportion of agricultural labourers increased considerably during the period 1961-71. Different authors put forward different explanations regarding the 1961-71 phenomenon among which the following explanation got special importance among the scholars working in this field. Thus, the decline of sharecropping partly due to land reform legislations and partly due to

TABLE 4

Distribution of Maiu Workers by Cultivators, Agl. Labourers and Other Workers, 1961-1981 : All India Rural & Urban

Census year	No. of cultivators	No of agl. labourers	No. of other workers	Total main workers
1	2	3	4	5
1961 : Total	95,548,714	31,332,978	55,660,910	182,542,602
	(52.34)	(17.17)	(30.49)	(100.00)
Male	63,685,692	17,158,617	44,102,091	124,946,400
	(50.97)	(13.73)	(35.30)	(100.00)
Female	31,863,022	14,174,361	11,558,819	57,596,202
	(55.32)	(24.61)	(20.07)	(100.00)
1971 : Total	75,093,398	47,046,217	52,882,997	175,022,612
	(42.91)	(26.88)	(30.21)	(100.00)
Male	65,896,233	31,272,626	46,919,955	144,088,814
	(45.73)	(21.71)	(32.56)	(100.00)

contd.......

1	2	3	4	5
Female	9,197,165 (29.73)	15,773,591 (50.99)	5,963,042 (19.28)	30,933,798 (100.00)
1981 : Total	91,394,401 (41.53)	55,371,405 (25.16)	73,316,725 (33.31)	220,082,531 (100.00)
Male	76,209,748 (44.77)	34,424,547 (19.77)	63,481,089 (36.46)	174,115,384 (100.00)
Female	15,184,653 (33.03)	20,946,858 (45.57)	9,835,636 (21.40)	45,967,147 (100.00)

Figures in parenthesis indicate percentage to total workers

Source : *Census of India 1981*. Series-1, INDIA, Paper-3 of 1981, Provisional Population
For Col. (5) P. 37 Totals : Workers and non-wo kers Registrar General
For Cols. (2)-(4), P-43 & Census Commission of India.

the technological innovations has led to a major decline in the proportion of the self-employed and a rise in the proportion of the wage employed in the rural working force (Bandyopadhyay 1975; Rudra and Newaj 1975; Raj 1976; Bardhan, P. 1976;). That is, eviction of tenants seemed to be a major cause in swelling the numbers of agricultural labourers in India during the period 1961-71.

Recent data collected by the census of India regarding the numbers of cultivators and agricultural labourers do not satisfy the above explanation unrestrictedly. Admittedly, the concept of cultivator used by the censuses is very broad and does not specify the actual composition of landless and landed groups within the cultivator category. But it is clear from the concept that the cultivators, by and large do not depend on wage earning as a principal occupation and they are something economically better off than the labouring class. Viewed from this angle higher rate of increase in the cultivator category in the recent decade may be considered an indication of upward mobility of agricultural population.

3. Labour Transfer

Theoretically it is believed that the process of economic development has been historically associated with the shift of labour force from less productive agricultural sector to more productive non-agricultural sector. The models of economic development for underdeveloped countries, in particular those of Nurkse (1953) and Lewis (1954), explicitly provide this. According to these models, the agricultural sector is typically the scene of serious under-employment or disguised unemployment. The marginal workers on most cultivating holdings are making little useful contribution to output. Therefore, the removal of marginal workers from their present occupations and their absorptions in work in the modern sectors, especially in capital construction, will promote development in the country in terms of its employment and output.

The aim of this section is to investigate in the Indian context the possibility or feasibility of intersectoral transfers of labour force on the basis of the change in the industrial distribution of working force over time and to compare them with similar changes in employment over time in each sector.

We do not have adequate data to examine the changes in the structure of labour force over time in different sectors of the economy. The National Sample Survey (NSS) which used to provide labour force data annually for some specific sectors in our economy was suspended in 1962 and thus annual data were not systematically available for all the sectors from this source. The data furnished by the Census of India at least cover the broad sectors of our economy for some specific points in time. But frequent changes in the concepts and definitions of workers and working force have placed very serious limitations on the use of Census data for inter-census comparisons of the number of workers or working force. However, the census returns could be used for the purpose of *relative* comparison, i.e., for finding out the proportion which the working force constitute of the total population in each sector at one census enumeration to another.

Presented in Table 5 are the results of analysis regarding the sectoral distribution of labour force, collated from the census data, for the period 1901-71.

It is seen from Table 5 that one can discover the tendency of working force increasing with time in the primary sector, decreasing with time in the secondary sector and the tendency of increasing with time in the tertiary sector. These changes imply that the secondary sector (industrial sector) has failed to catch the over-growing working force of our country, and therefore, working force both in the primary sector (agricultural sector) and tertiary sector (service sector) has been increasing over time.

In order to interpret the patterns described above more clearly, an examination of the relative movement of the additional working force in each census decade over various sectors of the economy may be useful. In this context, the data on economically active population in the Census of 1961 and 1971 may be better attempted instead of all the sets of census data beginning the year 1901. For the working force data of 1961 for the male workers of both rural and urban India have been made comparable with the fulltime workers of 1971 which has not been possible for female workers and also not been possible for other rounds of census data (see for details, Mahajan 1975). Thus, we may turn our attention to the working foıce data of 1961 and 1971 for the male workers to examine the phenomenon described above more meaningfully.

TABLE 5

Percentage distribution of working force (Male & Female in three broad sectors : All India, 1901-71)

Year	Primary sector (I+II+III)	Secondary sector (IV+V+VI)	Tertiary sector (VII+VIII+IX
(1)	(2)	(3)	(4)
1901	71.76	12.61	15.63
1911	74.86	11.13	14.01
1921	75.99	10.41	13.60
1931	74.75	10.21	15.04
1951	72.12	10.62	17.26
1961	72.28	11.70	16.02
1971	72.60	10.70	16.70

(I: Cultivators; II: Agricultural labourers; III: Mining, Quarying, Livestock, Forestry, Fishing, Hunting & Plantations and Allied Activities; IV : Household Industry; V: Manufacturing other than Household Industry; VI: Construction; VII : Trade and Commerce; VIII: Transport storage and communications; IX: Other services.

Sources: (a) *Census of India 1961*: Paper No 1 of 1962 Final population totals.

(b) *Census of India 1971* : Paper I of 1971 supplement provisional population totals.

TABLE 6

Sectrol distribntion of increase in male working force : All India 1961-71

Sectors	Percentage increase
(1)	(2)
Primary sector	81.3
Secondary sector :	6.6
(a) Household industry	—4.5
(b) Non-household industry	10.1
(c) Construction	0.9
Tertiary sector :	12.1
(a) Trade and Commerce	10.3
(b) Transport, Storage etc.	4.6
(c) Other services	—2.8
Increase in rural workers in primary sector as a percentage of increase in total rural workers	103.3

(Source : Mahajan, O.P., 1975, p. 40)

Table 6 shows that during the decade 1961-71, 81.3% of the total increase in male working force (2.97 crores) in the country as a whole has taken place in the primary sector as against 6.6% in the secondary, and 12.1% in the tertiary sector. It also follows from the table that the increase in the proportion of the primary sector has been caused mainly by decline in the proportions of workers engaged in the household industry and other services. Thus, it means that the industrial sector has no capacity to absorb the over-growing working force of our country and therefore a reverse shift from the more productive urban areas to the less productive rural areas is taking place, possibly among those workers whose relatives live in the rural areas for maintaining their livelihood in agricultural occupations. However, this needs further empirical investigation and analysis which is beyond the scope of census data.

4. Employment and Unemployment

In order to get further insight into the problem referred to above, we can discuss some changes in the structure of employment and unemployment of rural and urban India on the basis of NSS data which have been virtually the only source relating to employment and unemployment. It provides data both for rural and urban India at different points of time.

NSS estimates show that number of employed persons both in the rural and urban areas has been increasing during the period 1955 to 1972-73. Number of unemployed persons both in the rural and urban areas, however, shows no particular trend of increasing or decreasing over the time period (see for details, Chattopadhyay 1985).

Trends in employment and unemployment provided by the NSS data are consistent with evidences indicating a major break-through in agriculture during the sixties, In fact, some measures for land improvement, viz., expansion of irrigation, increased use of fertilizers and improved seeds, higher intensity of cropping etc., have taken place in the countryside in the 1960s. The immediate impact of these factors seems to have been increased employment and output in the rural sector. In this situation, it is quite natural that no shift of workers from agricultural sector to industrial sector will take place. Instead, the reverse shift from the industrial/urban sector to the agricultural/rural sector is expected to take place for the reason that the marginal rural migrants working in the urban areas will be attracted to go back to the villages for sharing work with their kith and kin.

5. Underemployment

We now proceed further into an analysis of underemployment data. For this purpose, we may consider data collected in some rounds of the NSS about hours of working time in the reference period which give some idea about the nature and extent of under-employment among the working persons of rural and urban India. Data are presented for four groups, defined in terms of weekly hours at work: 28 hours and less, 29 to 42 hours, 43 to 56 hours, and 57 and above hours. This kind of data is available only from the NSS particularly for the years 1955-65. No other data-sets are systematically and comprehensively available for this purpose.

Presented in Table 7 are the results of analysis based on the NSS data regarding underemployment. In the table, fully employed persons are defined to be persons with more than 43 hours of work in the reference week; underemployed persons are defined to be persons with less than 43 hours of work in the reference week. Here, the absolute numbers of the underemployed and the fully employed have been obtained by applying the percentages of different kinds of underemployment (provided by the NSS) to population estimates (see for details, Chattopadhyay 1985).

TABLE 7

Dimensions of Unemployment by Rural and Urban India Over Time (Based on NSS Estimates)

	Rural		Urban	
Year	Fully employed persons (in Mn.)	Under employed persons (in Mn.)	Fully employed persons (in Mn.)	Under employed persons (in Mn.)
1	2	3	4	5
1954-55	74.34	64.36	13.68	8.70
1955-56	66.53	62.44	14.45	8.56
1956-57	72.74	50.94	14.62	7.89
1958-59	70.96	66.17	15.81	8.34
1959-60	76.96	63.01	15.70	9.09
1960-61	89.43	55.51	—	—
1964-65	104.11	46.42	23.12	6.36

The results of analysis, presented in Table 7, reveal a decreasing tendency of underemployed persons and an increasing tendency of fully employed persons in both rural and urban India.

6. Surplus labour

With the kind of trend of underemployment in both rural and urban India presented above it will be interesting to assess how much of the underemployed persons can be removed from each sector without leading to any reduction in output. Following this approach, an estimate of removable surplus labour can be obtained from the aggregate data of the NSS for rural and urban separately which will indicate very roughly the amount of unproductive working persons for which new employment is needed to make them productive ones. The following methodology can be used for this purpose.

Let 'L' be the total number of workers for a particular time-point in a particular sector. L can be divided into two component parts say, L' and L'', where L' represents the number of fully employed workers (getting 43 hours or more of employment per week throughout the year) and L'' represents the number of underemployed workers (getting less than 43 hours of employment per week throughout the year). Now by considering the full employment norm described above, we can estimate the number of labourers who may be employed fully among L''. Let us call this L''.

With the help of the above three attributes (L, L' and L''), we can get the volume of surplus labour in the particular sector for a particular year.

Thus, surplus labour $S(L)=[L-(L+L'']$

Table 8 will give the results of estimation of surplus labour in rural and urban India separately for the period 1955-65.

Table 8 shows that surplus labour both in the rural and urban areas has been decreasing during the period 1955 to 1964-65. Although, the volume of surplus labour decreased in both the sectors in the 1960s compared to the 1950s, there

TABLE 8

Surplus labour in Rural and Urban India Over Time (Based on NSS data)

Year	*Rural* *No. of labourers per 100 (L)*			*Urban* *No. of labourers per 100 (L)*		
	Fully employed L′	Underemployed, but may be fully employed L″	Surplus labour S(L)= L—(L′+L″)	Fully employed L′	Underemployed, but may be fully employed L″	Surplus labour S(L) L—(L′+L″)
1	2	3	4	5	6	7
1954-55	53.70	24.86	21.44	60.47	23.90	15.63
1955-56	51.60	25.49	22.91	62.80	23.36	13.84
1956-57	58.92	24.20	16.88	65.71	22.53	11.76
1957-58	—	—	—	65.93	22.46	11.61
1958-59	51.56	24.98	23.46	64.82	22.80	12.38
1959-60	53.08	24.91	22.01	62.95	23.32	23.73
1960-61	60.92	23.81	15.27	—	—	—
1961-62	62.25	23.50	14.25	—	—	—
1964-65	69.26	21.29	9.45	78.07	17.12	4.81

is however, no indication of the non existence of the phenomenon of surplus labour in any of these two sectors. This once again indicate that both the sectors are relatively overburdened not only with the problem of acute unemployment but with the problem of surplus labour also.

7. Labour stagnation

The results of our analysis presented in the preceding sections venture to suggest that there is no justification of thinking of possibility of transfer of workers from agricultural to non-agricultural or from rural to urban areas. In fact, if such shift takes place, it will only add to the pool of jobless and surplus people in the non agricultural sector.

In this connection, we may refer to the results of a study, done by Khusro (1962), where it has been shown that unless the rate of capital formation in the non-agricultural sector is stepped up, the shift of workers from agriculture to non-agriculture will be economically useless. Khusro used data of non-agricultural investment and employment mainly for the period 1948-49 to 1950-51 and deducted the likely situation on the basis of these data upto 1976. These data indicated the increasing trend of urban unemployment in relation to the low rates of industrial growth vis-a-vis the faster rate of population growth in urban areas. The results of our analysis also suggest the conclusion drawn by Khusro, although the data have not been presented here in the same fashion as done by him.

From the above discussion, we may now conclude that both rural and urban India are overburdened with the problems of unemployment and surplus labour indicating thereby the henomenon of labour stagnation as both the sectors have failed to increase the employment opportunities satisfactorily among the growing labour force. In this situation, it is very difficult to believe that Indian economy is moving through such aprocess of economic development where inter-sector transfers of labour force, particularly from the agriculturalsector to the non-agricultural sector, are taking place to promote employment and output.

REFERENCES

Bandyopadhyay, N. (1975) : "Changing Forms of Agricultural Enterprise in West Bengal, *Economic and Political Weekly*, April 26.

Bordhan, K. (1977) : "Rural Employment, Wages and Labour Markets in India : A Survey of Research", *Economic and Political Weekly*, June 25, July 2 and 9.

Bardhan, P.K. (1976) : "Variations in Extent and Forms of Agricultural Tenancy", *Economic and Political Weekly*, May 26.

Billings, M.H. and Singh, A. (1969) : "Labour and Green Revolution: The Experience of Punjab," *Economic and Political Weekly*, Review of Agriculture, December 25.

Khan, A.S. (1974) : "The Impact of New Farm Technology on Structural Changes in Farm Organizations in the Punjab" *in Green Revolution : The Unfinished Task*, Minerva Associate Pvt. Ltd., Calcutta.

Raj, K.N. (1976) : "Trends in Rural Unemployment in India : An Analysis with Reference to conceptual and Measurement Problems", *Economic and Political Weekly*, Special Number, August.

Rudra, A. and Newaj, K. (1975) "Agrarian Transformation in a : District of West Bengal", *Economic and Political Weekly*, Review of Agriculture, March 29.

Visaria, P. (1976) : "Employment : A Trend Report", *Survey of Research in Economics*, vol. II, Macroeconomics, ICSSR, Allied Publishers, New Delhi.

Chattopadhyay, M. (1985) : *Conditions of Labour in Indian Agriculture : Apparent and Real*, K.P. Bagchi & Co., Calcutta.

Khusro, A.M. (1962) : *Economic Development with No Population Transfers : A Study in Demand for and Supply of Labour in the Non-Agricultural Sector of the Indian Economy.* Asia Publishing House, Bombay.

Lewis, W.A. (1954) : *Economic Development with Unlimited Supplies of Labour.* The Manchester School.

Mahajan, O.P. (1975) : "Economic Analysis of the Census of 1971 : Changes in the Structure of Employment in the Process of Economic Development of States" in : *Economic Analysis of the 1971 Census*, Popular Prakashan, Bombay.

Nurkse, R.)1953) : *Problems of Capital Formation in Under-Developed Countries.* Oxford.

2

Land Tenure and Distribution of Land

1. Propositions

There are two important questions relating to the share contact of farm land that have not been examined adequately in terms of empirical analysis. The two questions may be posed in the following words.

(1) It is quite well-known that concentration in the ownership of land is a characteristic feature of most agrarian economies. Those who have larger holdings generally lease out land to others, that is, the others to lease in, and therefore, the amount of operational holdings is different from that of ownership holding.

It this connection, some scholars working in this field have examined this question by assuming that in most agrarian economies, the transfer of land may take place unevenly to all types of sized-holdings. According to them, the concept of "optimal size of operational holdings" is not applicable in any situation in any agrarian economies[1]. They have, however, examined the "why not" question by identifying some relevant socio-economic factors in terms of theoretical models (e.g., Johnson 1950; Shapley and Shubik 1967; Day 1967 etc.). Some others have examined their significance in terms of empirical observations and analysis (e.g., Raj 1970; Bardhan and Srinivasan 1971, etc.). In the following, this aspect of the tenancy question is attempted to examine.

(2) The second question, as to whether all tenants are poor or not as compared to agricultural labourers, has not been fully taken care of in tenancy studies. Incidence of crop-sharing tenancy has been explained on the basis of factors such as wage rate, irrigation, production uncertainty, market conditions, availability of employment in non-agricultural scctor, etc. Among these several factors, the wage rate or net income of both agricultural labourer and tenant cultivator is believed to be one of the most important factors for explaining the incidence of tenancy. Next section attempts to study this relationship.

2. Ownership and Distribution of Land

Detailed data are available on the size, distribution of ownership and operational holdings in different zones as well as on the percentage of area leased in and leased out in different size-groups from the Farm Management Survey data. We have arranged the data in terms of the following notations for case of analysis.

Suppose 'F' is the total cultivated area for a particular farm. The farmer of this farm can lease out to others, say F_1 amount of land, out of F or he can lease in from others say F_2 amount of land and can add to F. After this give and take suppose the size of operational holding becomes F^1 (i.e., either $F\text{-}F_1=F^1$ or $F+F_2=F^1$ as the case may be). Obviously, F^1 must not be equal to F. Now, breaking F^1 into two componant parts, we get (a) amout of owned land after leasing out (if it happens) sayF^1, and (b) amount of leased in land, say $F^1{}_2$, then it follows that $F^1=F^1{}_1+F^1{}_2$.

Now, we may compute $F^1{}_1$ and F_2 for all the farms against each size-group. In Table 1, we have placed $F^1{}_1$ and against F^1 in the coloumns (3), (4) and (2) respectively for different size groups of different regions in India. We have taken the percentages of $F^1{}_1$ and $F^1{}_2$ to F^1 for indicating the variation of transfer of land over various size-groups.

It is found from Table 1 that the amount of transfer of land (F^1_2) in comparison to total operational holding (F^1) is more or less holding same for all the size groups in the regions of Uttar Pradesh, Andhra Pradesh, Orissa (1967-68) and Maharashtra (1967-68). Variation of F^1_2 according to farm-size groups is noticed in the region of Punjab, Tamil Nadu, Madhya Pradesh, Assam and Orissa (1968-70).

Thus, the basic assumption that the transfer of land does not equally take place over various size-groups, does not unequivocally satisfy the figures for all regions under study. The assumption is of course valid in case of some regions, but the reverse is also true in some other regions. It is seen that in some agriculturally better-endowed regions (in terms of its irrigation, cropping pattern, fertility of soil etc.), such as Punjab, Tamil Nadu, Madhya Pradesh etc., the transfer of land varies considerably with farm size, but the same is not true for the same type of regions like Uttar Pradesh, Andhra Pradesh etc. In the agriculturally less progressive regions, the same conclusion may be drawn. Hence the explanation that a region with better irrigation, higher fertility of soil, mechanised farming etc., will tend to have a larger incidence of cropsharing tenancy is not acceptable unambiguously. Apart from these factors, it is argued that some external factors like market, production uncertainty, wages or earnings etc, may explain the incidence of tenancy more meaningfully. Since wages or earnings of labourers of share-croppers are mainly guided by the influence of market and production uncertainty, it may be useful to examine the incidence of tenancy in terms of their income status rather than to consider the other elements of the production process, the following section attempts to examine this.

3. Types of Tenure and Income Distribution

In most of the tenancy studies it is assumed that the wage earnings of agricultural labourer must be equal to the net income of sharecropper[2]. If wage rate of agricultural labourer is greater than the daily earning of share-cropper, then the whole system of share-cropping will be ruled out simply because working as agricultural labourer will be more remunerative. That is, share-cropping business will turn into the occupation of wage earning and *vice-versa.* Therefore in most of the tenancy studies, the uneven distribution of operational holdings according to farm size is explained by factors like irrigation, fertility of soil, cropping pattern etc., and not by the wage rate. Bardhan and Srinivasan (1971), however observed that the incidence of sharecropping, cannot be explained only by the factors like irrigation and according to them, the wage rate also plays a positive role for explaining it[3]. In this context, the hypothesis that a region with a higher wage rate will tend to have a larger incidence of cropsharing tenancy, needs proper investigation. Before examining this hypothesis, it is necessary to take a judgement about the relative economic position of three categories of agricultural population, namely, owner cultivator, sharecropper and agricultural labou-

TABLE 3

Items	States
1. Net income of owner cultivator > net income of share-cropper	UP (1966-67, 1968-69), Punjab (1969-70), Orissa (1969-70) AP (1967-70), TN (1967-70), MP (1962-65), Kerala (1962-65)
2. Net income of owner cultivator < net income (of share-cropper)	UP (1967-68), Maharashtra (1967-68), Assam (1968-69)
3. Net income of owner cultivator = net income of share-cropper	Nil
4. Net income of share-cropper > hired labour charges	UP (1966-69), Punjab (1967-68), Orissa (1967-70), Assam 1968-69)
5. Net income of share-cropper < hired labour chaiges	Punjab (1968-70), AP (1967-70), TN (1967-70), MP (1962-65), Kerala (1962-65), Maharashtra (1967-68)
6. Net income of share-cropper = hired labour charges	Nil

rer. We have presented our results in Table 2 which demonstrate the net income *per hectare* of the three groups of agricultural population, namely owner cultivator, sharecropper or owner-cum-tenant cultivator, as well as the income of agricultural labour from *per hectare* of land respectively. The abstract picture of our findings can be perceived from Table 3.

It is found from Tables 2 & 3 that in majority of the cases (14 out of 17), the net income of owner cultivator is greater than the net income of share-cropper. It is not at all surprising, because, as the 'rent' factor is included for hiring the land, the cost of cultivation per hectare of share-cropping land is greater than the owner cultivating land. The interesting feature, however, is that some regions at some points of time (3 cases) do not follow this pattern. In these cases, output per hectare of sharecropping land is considerably greater than the output of owned cultivated land and the cost of cultivation for sharecropping land is also not proportionately greater than the owned cultivated land. Therefore, the economic position of sharecroppers seems to be a little better than that of owner cultivators in these few regions. From these observations it would not be misleading to conclude that all tenants are not poor peasants; leasing in land sometimes becomes an economic enterprise leading to better economic status.

Now, if we compare the net income of share-cropper with the wage income of agricultural labourer, it is found from the same tables that in 8 cases out of 17, the net income of share-cropper is greater than the wage income of agricultural labourer per hectare of land. And, in the remaining 9 cases, the former is lower than the latter. Thus, our findings, in no cases, support the basic assumption that net income of sharecropping must be equal to the wage rate of agricultural labourer. However, the question arises : why in some regions the share-croppers under low income status compared to agricultural labourer do not shift their way of life away from sharecropping to agricutural labour? The most plausible explanation for this phenomenon may be that since the employment and income of a share-cropper is rather guaranteed for the whole year as compared to the agricultural labourer, the former gets a psychological boost and again, the sharecropper enjoys a relative freedom compared to the labourer. Therefore, he sticks himself to the share-cropping occupation for a minimum economic security. On the other hand, in some regions the agricultural labourers with lower wage rate cannot shift their way of life to the share-cropping occupation due to lack of initial money for purchasing working capital in the forms of bullocks of ploughs, seeds, manures etc. For these twin reasons, both the occu pations, share-cropping and wage earning, exist side by side in Indian agriculture.

4. Incidence of Sharecropping and Wage Rate

We may now turn to examine the relationship between wage rate and incidence of share-cropping. For this purpose an attempt has been made to compare the variations of leased in land with the wage rate for those regions where data are available. Presented in Table 4 are the results of our analysis.

If we rank the 8 States according to the incidence of daily wage rate of agricultural labourer, the positions of the States in descending order are as follows :-

1. Punjab, 2. Assam, 3. Uttar Pradesh, 4. Orissa, 5. Tamil Nadu, 6. Maharashtra, 7. Andhra Pradesh, 8. Madhya Pradesh.

Similarly, if we roughly do rank the 8 States according to the incidence of tenancy, their positions in descending order are as follows:-

1. Punjab, 2. Orissa, 3. Madhya Pradesh, 4. Assam, 5. Tamil Nadu, 6. Maharashtra, 7. Andhra Pradesh, 8. Uttar Pradesh.

From these comparisons it is seen that Punjab, Tamil Nadu, Maharashtra and Andhra Pradesh maintain the same position for both the characteristics. For these four States we can infer that the incidence of tenancy is positively related to the wage rate. But for the rest of the regions, viz., Assam, Uttar Pradesh, Orissa and Madhya Pradesh, we cannot make such inference. Hence, the hypothesis that a region with a higher wage rate will tend to have a larger incidence of

TABLE 4

State		Percentage of leased in land to total cultivated area	Daily wage rate of agricultural labourer (Rs.)
(1)		(2)	(3)
Uttar Pradesh :	1966-67	1.68	2.66
	1967-68	0.42	2.89
	1968-69	0.74	2.93
Punjab :	1967-68	25.59	4.64
	1968-69	36.80	4.80
	1969-70	36.73	5.68
Orissa :	1967-68	8.86	2.44
	1968-69	43.20	2.59
	1969-70	37.49	2.72
Andhra Pradesh :	1967-68	2.92	2.04
	1968-69	1.08	2.00
	1969-70	1.35	2.01
Tamil Nadu :	1967-68	22.06	2.42**
Madhya Pradesh :	1962-65	34.61	1.46*
Maharashtra :	1967-68	5.34	2.28**
Assam :	1968-69	28.69	3.69

*relates to 1964-65

**relates to 1970-71

Source : Col. (2) : *Studies in Economics of Farm Management*, Govt. of India.

Col. (3) : *Agricultural Wages in India*, Govt. of India.

cropshaiing tenancy is partly supported by our data. It seems to us that no particular factor by and large can explain the incidence of share-cropping in Indian agriculture.

We may now conclude that the 'wage' based explanation may possibly be the one explanation for rise and fall of cropsharing tenancy where it holds, but certainly not one that is universally valid or which takes precedence over various explanations that have been advanced by other research workers. There are possibly various reasons why cropsharing tenancy behave in the fashion described above. The different factors that have been put forward by different authors are not really rivals. They very often operate in quite different socio-economic setups and in different geographical contexts resulting thereby in the cropsharing tenancy declines or rises.

1. In this context, K.N. Raj (1970) writes : "There is no reason however why the units in which production is organized should correspond in size to the ownership holdings. The optimal size of the operational holdings (i.e., the actual units of production and management) will depend on the other factor endowments and the technical conditions of production. One would expect those who have holdings above this size to lease out land, the others to base in, and the pattern of operational holdings to be correspondingly different from that of ownership." (p. 1)

"That transfers do take place is not in question, but there is reason to doubt the extent to which the pattern of distribution of land is modified by such transfers in the less developed agrarian economies." (p.1)

2. In this connection, the theoretical models and the properties of equilibrium, put forward by K.N. Raj (1970) and Bardhan & Srinivasan (1971) may be seen.

3. Bardhan and Srinivasan (1971) writes : "Our main independent variables are W, the agricultural wage rate, and I, the percentage of operated area irrigated. Needless to say, in the real world S (Percentage of operated area leased in under

crop-sharing tenancy) will depend on various other economic and particularly non economic factors. But our purpose is only to test if we have been able to identify at least two major economic factors explaining variations in the incidence of share-cropping and if the signs of the regression coefficients for these two variables are as expected from our comparative-state analysis" (parenthesis added) (p. 57).

"......the regression coefficients for W and I are significant (at 5 perc ent level) and of expected positive signs." (p. 58)

TABLE 1

Farm size groups (hect.)	Total operational area (hect) F^1	Cultivated area under $F^1{}_1$	$F^1{}_2$	% of col. (4) to col. (2)	% of col. (3) to col. (2)
1	2	3	4	5	6
		Uttar Pradesh (Muzaffarnagar) : 1966-67			
Below—2.87	58.99	56.55	2.44	4.14	95.86
2.88—4.71	116.19	112.42	3.77	3.24	96.76
4.72—6.96	167.01	163.16	3.85	2.31	97.69
6.97—10.65	261.05	253.49	7.56	2.90	97.10
10.66 & above	365.17	365.17	0.00	0.00	100.00
All	968.41	950.79	17.62	1.82	98.18
		Uttar Pradesh (Muzaffarnagar) : 1967-68			
Below 2.87	59.26	56.49	2.77	4.46	95.54
2.88 — 4.71	116.69	116.05	0.64	0.55	99.45
4.72 — 6.96	168.46	167.71	0.75	0.44	99.55
6.97 —10.65	249.65	249.65	0.00	0.00	100.00
10.66 & above	381.32	381.32	0.00	0.00	100.00
All	975.37	971.22	4.15	0.42	99.58

1	2	3	4	5	6
Uttar Pradesh (Muzaffarnagar) : 1968-69					
Below 2.87	63.57	60.79	2.78	4.37	95.63
2.88 — 4.71	107.87	107.87	0.00	0.00	100.00
4.72— 6.96	170.19	168.46	1.73	1.02	98.98
6.97—10.65	250.84	248.06	2.78	1.11	98.89
10.66 & above	394.78	394.78	0.00	0.00	100.00
All	987.25	979.96	7.29	0.74	99.26
Punjab (Ferozepur) : 1967-68					
Below 6	142.02	107.42	34.60	24.36	75.64
6— 9	201.03	168.46	32.57	16.20	83.80
9— 14	458.32	284.30	174.02	37.97	62.03
14— 24	669.91	449.97	219.94	32.83	67.17
24 & above	424.97	400.85	24.12	14.63	85.37
All	1896.25	1411.00	485.25	25.59	74.41
Punjab (Ferozepur) : 1968-69					
Below 6	153.12	117.51	35.61	23.26	76.74
6— 9	204.06	123.10	80.96	39.67	60.33
9— 14	428.99	237.82	191.17	44.56	55.44
14— 24	670.89	394.44	276.45	41.21	58.79
24 & above	388.77	293.75	95.03	24.44	75.56
All	1845.83	1166.62	679.21	36.80	63.20
Punjab (Ferozepur) : 1969-70					
Below 6	142.93	92.42	50.51	35.34	64.66
6— 9	256.42	147.21	109.21	42.59	57.41
9—14	348.02	238.41	109.61	31.50	68.50
14—24	636.76	354.72	282.04	44.29	55.71
24 & above	470.67	340.72	129.95	27.61	72.39
All	1854.80	1173.48	681.32	36.73	63.27

(1)	(2)	(3)	(4)	(5)	(6)
Orissa (Cuttack) : 1967-68					
0 —0.81	12.50	11.22	1.28	10.24	89.76
0.82—1.32	39.49	34.76	4.73	11.98	88.02
1.33—2.03	49.68	62.55	7.13	14.35	85.65
2.04—3.64	102.51	91.87	10.64	10.16	89.84
3.65&Above	76.03	74.74	1.29	1.69	98.31
All	280.21	255.14	25.07	8.86	91.14
Orissa (Cuttack) : 1968-69					
0 —0.81	13.92	7.04	6.88	49.42	50.58
0.82—1.32	37.08	18.39	18.69	50.40	49.60
1.33—3.03	57.44	26.30	31.14	54.21	45.79
2.04—3.64	101.79	67.71	34.08	33.48	66.52
3.65&Above	60.11	34.10	26.10	43.27	56.70
All	270.34	153.54	116.80	43.20	56.80
Orissa (Cuttack) : 1969-70					
0 —0.81	11.07	8.30	2.77	25.02	74.98
0.82—1.32	38.02	17.92	20.10	53.87	47.13
1.33—2.03	62.12	28.62	33.50	53.93	46.07
2.04—3.64	105.85	73.90	31.95	30.18	69.82
3.65—& Above	47.43	36.58	10.85	22.87	77.13
All	264.69	165.52	99.17	37.49	62.51
Andhra Pradesh (Cuddaph) : 1967-68					
Below 1.62	27.72	24.10	3.62	13.06	86.94
1.63—3.23	79.07	79.07	0.00	0.00	100.00
3.24—6.07	155.92	144.44	11.48	7.36	92.64

(1)	(2)	(3)	(4)	(5)	(6)
6.08—11.33	286.52	278.71	7.81	2.73	97.27
11.34 & Above	320.14	317.69	2.45	0.77	99.23
All	869.37	844.01	25.36	2.92	97.08
Andhra Pradesh (Cuddapah) : 1968-69					
Below 1.62	27.91	27.55	0.36	1.29	98.71
1.63— 3.23	78.97	77.96	1.01	1.28	98.72
3.24— 6.07	150.63	148.06	2.57	1.71	98.29
6.08—11.33	283.97	283.37	0.60	0.21	99.79
11.34& Above	333.61	328.74	4.87	1.46	98.54
All	875.09	865.68	9.41	1.08	98.92
Andhra Pradesh (Cuddapah) : 1969-70					
Below 1.62	27.43	27.31	0.12	0.44	99.56
1.63— 3.23	79.77	77.82	1.95	2.44	97.56
3.24— 6.07	155.85	153.48	2.37	1.52	98.48
6.08—11.33	286.83	284.37	2.46	0.86	99.14
11.34 & Above	319.72	314.85	4.87	1.52	98.48
All	869.60	857.83	11.77	1.35	98.65
Tamil Nadu (Thanjavur) : 1967—70					
Up to 1.16	23.31	12.66	10.65	45.69	54.31
1.17—2.02	58.45	31.14	27.31	46.72	53.28
2.03—3.05	85.14	48.22	36.92	43.36	56.64
3.06—5.71	108.92	90.23	18.69	17.16	82.84
Above 5.71	187.03	178.50	8.53	4.56	95.44
All	462.85	360.75	102.10	22.06	77.94

(1)	(2)	(3)	(4)	(5)	(6)
		Madhya Pradesh (Rajpur) : 1962-65			
0.01—1.00	13.88	13.34	0.54	4.68	95.32
1.01—2.00	35.29	29.64	5.65	15.83	84.17
2.01—4.00	61.93	39.02	22.91	37.45	62.55
4.01—6.00	59.93	47.94	11.99	20.21	79.79
6.01& Above	168.05	94.11	73.94	44.39	55.61
All	339.08	224.05	115.03	34.61	65.39
		Maharashtra (Ahmednagar) : 1967-68			
4.24 & below	82.59	80.95	1.64	1.98	98.02
4.25— 7.07	159.01	155.77	3.24	2.04	97.96
7.08—10.71	220.48	201.49	18.99	8.63	91.37
10.72—16.38	353.63	319.72	33.91	9.58	90.42
16.38& Above	481.36	456.20	25.16	5.23	94.77
All	1297.07	121.13	482.94	5.34	94.66
		Assam (Nowgong) : 1968-69			
0.01—1.82	40.44	33.12	7.32	8.10	91.90
1.83—2.43	63.39	32.77	30.62	48.31	51.69
2.44—3.24	94.41	78.07	16.34	17.31	82.69
3.25—4.45	113.11	82.57	30.57	27.02	72.98
Above 4.45	171.86	118.08	53.78	31.29	68.71
All	483.24	344.61	138.63	28.69	71.31

Source : *Studies in Economics of Farm Management*, Directorate of Economics and Statistics, Government of India.

TABLE 2

Type of tenancy	Output per hect. (Rs.)	Input per hect. (Rs.)	Net income per hect. (Rs.)	Hired labour charges per hect. (Rs.)
(1)	(3)	(3)	(4)	(5)
Uttar Pradesh (Muzaffarnagar) : 1966-67				
Fully owned	2749.00	1528.00	1221.00	
Partially or wholly tenanted	2876.00	1792.00	1084.00	
All	2754.00	1539.00	1215.00	99.00
Uttar Pradesh (Muzaffarnagar) : 1967-68				
Fully owned	3866.00	1691.00	2175.00	
Partially or wholly tenanted	4386.00	2014.00	2372.00	
All	3877.00	1698.00	2179.00	97.00
Fully owned	3234.00	1746.00	1488.00	
Partially or wholly tenanted	2942.00	1736.00	1206.00	
All	3223.00	1716.00	1477.00	93.00
Punjab (Ferozepur) : 1968-69				
Fully owned	1799.20	1495.56	9303.64	
Owner-cum-tenant	1799.20	1544.86	254.34	
All	1799.20	1544.86	254.34	200.53
Punjab (Ferozepur) : 1968-69				
Fully owned	1687.06	1544.92	142.14	
Owner-cum-tenant	1687.06	1596.22	90.84	
All	1687.06	1596.22	90.84	181.60

(1)	(2)	(3)	(4)	(5)
		Punjab (Ferozepur) : 1969-70		
Fully owned	2009.44	1713.07	296.37	
Owner-cum-tenant	2009.44	1787.06	222.33	
All	2009.44	1787.06	222.38	238.82
		Orissa (Cuttack) : 1967-68		
Fully owned	1321.04	766.48	554.56	
Owner-cum-tenant	1212.91	890.30	322.61	
All	1298.99	801.08	497.91	112.10
Fully owned	1423.42	983.02	440.40	
Owner-cum-tenant	1473.14	1130.88	342.26	
All	1445.67	1049.20	396.47	147.01
		Orissa (Cuttack) : 1969-70		
Fully owned	1314.59	831.45	483.14	
Owner-cum-tenant	1394.98	1218.61	176.37	
All	1344-78	1043.35	301.43	126.82
		Andhra Pradesh (Cuddapah) : 1967-68		
Fully owned	946.41	974.77	—28.36	
Owner-cum-tenant	946.41	981.56	—35.15	
All	946.41	981.56	—35.15	117.98
		Andhra Pradesh (Cuddapah) : 1968.69		
Fully owned	798.09	725.96	72.13	
Owner-cum-tenant	798.09	734.08	64.01	
All	798.09	734.08	64.01	108.48
		Andhra Pradesh (Cuddapah) : 1969-70		
Fully owned	818.69	874.73	—56.04	
Owner-cum-tenant	818.69	878.10	—59.41	
All	818.69	878.10	—59.41	110.55
		Tamil Nadu (Thanjavur) : 1967-70		
Fully owned	1791.90	1593.70	198.20	373.04
Fully tenanted	1788.16	1601.28	186.88	375.48

(1)	(2)	(3)	(4)	(5)
Owner-cum-tenants	1985.47	1804.36	181.11	466.86
All	1840.35	1648.14	192.21	397.28
Madhya Pradesh (Raipur) : 1962-65				
Fully owned	431.78	330.24	101.54	
Owner-cum-tenant	431.78	341.95	89.83	
All	431.78	341.95	89.93	114.84
Kerala (Alleppey & Quilon) : 1962-65				
Fully owned	1033.62	1003.95	29.67	
Owner-cum-tenant	1033.62	1065.54	—31.92	
All	1033.62	1065.54	—31.92	33.12
Maharashtra (Ahmednagar) : 1967-68				
Fully owned	799.53	761.67	37.86	
Owner-cum-tenant	799.53	775.50	24.03	
All	799.53	775.50	24.03	64.32
Assam (Nowgong) : 1968-69				
Fully owned	1215.00	1119.00	96.00	
Fully tenanted	1330.00	1136.00	194.00	
Owner-cum-tenant	1208.00	1103.00	105.00	
All	1215.53	1115.00	100.53	79.71

Source : *Studies in Economics of Farm Management*, Directorate of Economics and Statistics, Government of India.

REFERENCES

Bardhan, P.K.
(1970) "Trends in land relations : A note, "*Economic and Political Weekly*, Annual No. January.

Bardhan, P.K. and Srinivasan, T.N.
(1971) "Cropsharing tenancy in agriculture : A theoretical and empirical analysis," *The American Economic Review*, March.

Day, R.H.
(1967) "The economics of technological change and the demise of the sharecropper," *The American Economic Review*, June.

Johnson, D.G.
(1950) "Resource allocation under share contacts," *The Journal of Political Economy*, April.

Raj, K.N.
(1970) "Ownership and distribution of land," *Indian Economic Review*, April (New series).

Shapley, S. and Shubik, M.
(1967) "Ownership and the production function", *The Quarterly Journal of Economics*, February.

3

Modes of Cultivation and Rates of Return

1. Introduction

Broadly speaking, over the sixties, a number of major changes has taken place in agriculture. From our present empirical understanding of the employment situation in particular, it appears that :

(i) Intensity of employment in agriculture has gone up during recent times for all the regions at any rate compared to pre-sixties;

(ii) Intensity of employment has gone up significantly since mid-sixties in the green-revolution areas compared to non-revolution areas;

(iii) Proportion of hired labour has gone up significantly for all the regions, espeacally in the green-revolution areas;

(iv) Share-cropping tenancy has declined and owner cultivation with hired labour has gone up;

(v) A sharp rise in the number of agricultural labourers has taken place along with the decline in self-employed non-agricultural workers.

From these one would expect to find different modes of production in recent times as compared to pre-sixties, at least for those regions where these changes are more prominent.

Mode of production in agriculture means, above all, the basis of farming which is primarily institutional in character, broadly embodies two types of farming—owner and tenant cultivation—having two techniques of farm production—hired labour any family labour-based. Thus, the agrarian structure of production relations has three-tier composition—owner cultivation,tenant cultivation and landless labourer. This three-tier structural composition of production relations indicate the different systems of farming—peasant, capitalist, seni-feudlal etc.—depending upon thc relations with each other in the production process through labour hiring and land leasing.[2]

Viewed from this angle, capitalist development in agriculture means, above all, the radical change in the technique of production, that is, virtual elimination of a particular technique and its replacement by another technique of production. Higher intensity of employment of hired labour, in this context, is regarded as the principal manifestation of agricultural capitalism. Lonin thus, wrote: "hired labour is the chief sign and indicator of capitalism in agriculture" (*Collected Works*, Vol. 22, P. 101) and again, "the data on hired labour are far more indicative and offer better proof of capitalism in agriculture" (*ibid*, P. 32). The principal distinguishing feature of this development is that "the ownership of means of production is clearly divorced from the performance of labour. The unity between these two is the chief feature of feudal production relations"...(Sen 1962).

Now, let us look at the empirical situation at the use of different kind of labourers by different categories of farmers, for understanding the modes of production in agriculture in an extensive manner.

2. Theoretical Approach

Corresponding to the three distinct modes of labour employment (family workers, casual workers, permanent servants) in an agarian economy, it has been argued that there may be distinct (market) or imputed wage rates at any point of the time and distinct process of determination with distinct sets of supply and demand curves under distinct set of modes of production. Subsequently, A.K. Sen (1962) formalised this as the "dual labour market hypothesis", with capitalist farms using hired labour at market wage higher than the value of family labour used on peasant farms. To examine this phenomenon, let us present some results pertaining to the incomes of different kinds of labourers under different modes of cultivation. Different modes of cultivation have different classificatory norms which may be defined as follows:

Mode 1: Fully family labour and own bullock labour based owner cultivating farms;

Mode 2: Fully family labour and own bullock labour based tenant cultivating farms.

Mode 3: Fully family labour and fully hired bullock labour based owner cultivating farms.

Mode 4: Fully family labour and fully hired bullock labour based tenant cultivating farms.

Mode 5: Fully hired casual labour and own bullock labour based owner cultivating farmsi.

Mode 6: Fully hired casual labour and own bullock labour based tenant cultivating farms.

Mode 7: Fully hired casual labour and hired bullock labour based owner cultivating farms.

Mode 8: Fully hired casual labour and hired bullock labour based tenant cultivating farms.

Mode 9: Fully permanent farm servant and own bullock labour based owner cultivating farms.

Mode 10: Fully permanent farm servant and own bullock labour based tenant cultivating farm.

Mode 11: Fully permanent farm servant and fully hired bullock labour based owner cultivating farms.

Mode 12: Fully permanent farm servant and fully hired bullock labour based tenant owner cultivating farms.

We may compare the incomes of different kinds of labourers (family workers, casual workers, and permanent farm servants) vis-a-vis the incomes of different categories of farms (small, medium, and large farms) under these twelve sets of modes of cultivation. The list of the modes of cultivation can be extended by introducing other new variables of cost of production. But let us restrict our analysis within these twelve sets which are mainly the basis of farming in Indian agriculture. The results of the types we have made, are of course, crude and hypothetical, being based on a number of assumptions which may be described as follows :

(i) Let the gross value of output from crop production (i.e. all crops and their by-products) be (Rs.) X. Let material cost for crop production be (Rs.) M, where

$$M=S+F+L+P+D$$

S=Value of seeds

F=Values of manures and fertilizers

I=Irrigation charges

P=Value of insecticides and pesticides

D=Depreciation

Value added V is then

$$V=X-M$$

Let L be the total human labour days employed for crop production

Then the human labour income per day can be obtained as:

$$W=\frac{V}{L}$$

If L indicates the employment of fully family labour, then W indicates the family labour income per day; if L indicates the employment of fully hired casual labour, then W indicates the casual labour income per day; and similarly W may indicate the income of permanent servant per day.

(ii) The incomes of different kinds of labourers can also be obtained in the purely tenant operated farms with the same procedure, provided the share of output (X) and share of costs of tenant are obtained. For sake of simplicity, we, in this exercise, have assumed the fifty per cent share of output with no cost sharing for each tenant operated farm.

(iii) Costs of hired bullock labour have been subtracted from X in the cases where the purely owner operated farms are hired bullock labour based. In the case of tenant operated farms, it has been deducted from the share of output, received by a tenant.

(iv) Needless to say, the incomes of different categories of farmers under different modes of cultivation may be obtained from the attributes X, V, W and B (hired bullock labour charges) as the case may be.

We have used aggregate data of Farm Management Surveys for the agricultural years beginning 1962 for some regions in India to compare the incomes of different kinds of labourers vis-a-vis the incomes of farms under different modes of cultivation. It is not our purpose in this study to examine the very many propositions that have been forwarded by different research workers for analysing the modes of production in agriculture. Our venture in this exercise only helps to understand the probable economic conditions of agricultural labourers under probable elements of change in the modes of production in Indian agriculture. To examine this, let us turn to the imputed results of our analysis.

2. *Rates of Return Under Different Types of Farming*

2.1 Family Labour based farms

Comparison of family labour income with wage labour income per day under four distinct modes of cultivation reveal (Table 14) the following tendencies:

TABLE 1A

Family labour income per day under fully family labour based farming for different regions (farm size groups as a whole)

Modes of cultivation	Rcgions with a agricultural years Assam 1968-69	Punjab 1969-70	UP 1968-69	AP 1969-70	Tamil Nadu 1969-70	Kerala 1962-65	W.B. 1972-73	Orissa 1969-70
(1)	(2)	(3)	(4)	(5)	(6)	(7)	(8)	(9)
M_1 : Fully family labour and own bullock labour based owner cultivating farms	11.97 (4.05)	16.01 (5.68)	28.95 (2.93)	7.70 (2.01)	9.08 (3.05)	5.68 (2.43)	16.79 (3.29)	7.63 (2.72)
M_2 : Fully family labour and own bullock labour based tenant cultivating farms	5.37 (4.05)	4.42 (5.68)	12.34 (2.93)	2.22 (2.01)	3.50 (3.05)	2.16 (2.43)	6.90 (3.29)	3.22 (2.72)
M_3 : Fully family abour and fully hired bullock labour based owner cultivating farms	10.85 (4.05)	14.25 (5.68)	24.99 (2.93)	6.88 (2.01)	8.50 (3.05)	5.31 (2.43)	NA	6.11 (2.72)
M_4 : Fully family labour and fully hired bullock labour based tenant cultivating farms	4.25 (4.05)	2.66 (5.68)	8.38 (2.93)	1.41 (2.01)	2.92 (3.05)	1.79 (2.43)	NA	1.70 (2.72)

(Figures in parenthesis indicate the market wage rates of casual labour)

(a) Family labour income per day is much higher than of wage labour income in the fully owner operated family labour based farms (M_1), even in the fully owner operated hired bullock labour based farms (M_3). This is true for all the regions under study, although regional variations of incomes are prominently noticed.

(b) Family labour income per day is also higher than of wage labour income in the fully tenant operated family labour based farms (M_2) for all the regions except for Punjab and Kerala. But this is not true in the case of fully tenant operated hired bullock labour based farms (M_4) for the majority of the regions under study.

TABLE 1 B

Family labour income per day under fully family labour based farming by farm size groups and regions

Farm size groups (in hectare)	Modes of cultivation M_1	M_2	M_3	M_4
(1)	(2)	(3)	(4)	(5)
		Assam: 1968—69		
0.01—1.82	11.55	5.17	10.39	4.01
1.83—2.43	12.36	5.57	11.15	4.35
2.44—3.24	11.21	4.98	10.12	3.89
3.25—4.45	12.76	5.75	11.63	4.62
Above 4.45	11.95	5.37	10.85	4.28
		Punjab: 1969-70		
Below 6	12.54	3.74	11.32	2.51
6—9	15.32	4.55	14.07	3.30
9—14	17.42	4.86	15.75	3.19
14—24	17.39	4.67	15.07	2.36
24 & above	25.89	8.53	22.84	5.48

1	2	3	4	5
Uttar Pradesh: 1968—69				
Below 2.87	20.93	8.98	16.40	4.45
2.88—4.71	26.56	11.27	22.37	7.08
4.72—6.96	26.15	11.22	22.38	7.45
6.97—10.65	33.20	14.19	28.92	10.02
10.66 & above	30.89	13.12	27.72	9.96
Andhra Pradesh: 1969-70				
Below 1.62	4.69	1.53	3.77	0.60
1.63—3.22	6.25	1.61	5.09	0.46
3.23-6.07	7.02	2.14	6.09	1.21
6.08—11.33	6.96	1.81	6.28	1.13
11.34 & above	9.79	2.96	8.99	2.17
Orissa: 1969-70				
0.01—0.81	7.62	3.04	6.10	1.52
0.82—1.32	8.01	3.30	6.44	1.66
1.33—2.03	7.80	3.34	6.30	1.84
2.04—3.64	7.22	3.04	5.63	1.46
3.65 & above	8.02	3.44	6.79	2.20
West Bengal: 1972-73				
0.01—0-50	14.31	6.08	N A	N A
0.51—1.00	14.89	6.30	,,	,,
1.01—1.50	15.76	6.50	,,	,,
1.51—2.00	18.26	7.60	,,	,,
2.01-3.00	20.58	8.82	,,	,,
3.01—4.00	16.38	5.98	,,	,,
4.01—6.00	——	——	,,	,,
6.01 & above	21.08	7.87	,,	,,
Tamil Nadu : 1969-70				
Up to 1.16	7.52	2.84	6.98	2.31

1		2	3	4	5
1.17—2.02	7.67	2.91	6.94	2.17
2.03—3.05	8.91	3.53	8.26	2.87
3.06—5.71	8.02	3.09	7.45	2.52
Above 5.71	11.03	4.23	10.57	3.77
		Kerala : 1962-65		
0—1.0	3.97	1.44	3.85	1.33
1.0—2.5	3.40	1.29	3.17	1.06
2.5—5.0	4.96	1.94	4.54	1.53
5.0—10.0	5.69	2.26	5.30	1.87
10.0—15.0	5.56	2.10	5.16	1.70
15.0—25.0	6.03	2.29	5.70	1.96
Above 25.0	7.76	2.91	7.39	2.53

Distribution of family labour income per day under different farm size groups also reveals the same tendencies, as noted under (a) & (b) (Table 1B). It also indicates that as farm size increases, family labour income per day by and large increases and this is true for all the regions under study. These results show that the imputed income of family labour to the small or the subsistence farms are not equal to the market wage rate but somewhat higher than that. Hence, the dual labour market hypothesis advanced by Sen, does not hold true in the majority of regions under study.

2.2 Labour based farms

Let us now turn to compare the incomes of purely hired labour based farms under different categories mentioned under M_5 to M_{12}. Presented in Table 2 are the results of our analysis in relation to purely hired casual labour based farms

It is seen from Table 2 that the income per unit of land is always positive in the fully owner operated-hired casual labour based farms. The same feature is discerned if these farms use fully hired bullock labour instead of fully or partly owned bullock labour. But, on the other hand, the income per unit of land in the fully tenant operated-hired casual labour

TABLE 2

Income per unit of land under purely hired casual labour based farming for different regions (farm size groups as whole)

Modes of cultivation	Regions with age. years							
	Assam (1968-69	Punjab 1969-70	UP 1968-69	AP 1969-70	Tamil Nadu 1969-70	Kerala 1962-65	WB 1972-73	Orissa 1969-70
(1)	(2)	(3)	(4)	(5)	(6)	(7)	(8)	(9)
M_5 : Fully hired casual labour and own bullock labour based owner cultivating farms	729.16	895.44	2523.79	125.23	1162.50	477.32	2723.72	148.32
M_6 ; Fully hired casual labour and own bullock labour based tenant cultivating farms	121.41	-109.28	912.29	15.89	86.67	-39.49	727.57	75.93
M_7 : Fully hired casual labour and hired bullock labour based owner cultivating farms	625.80	742.59	2139.79	364.36	1051.72	422.31	NA	517.30
M_8 : Fully hired casual labour and hired bullock labour based tenant cultivating farms	17.65	-262.13	528.29	-44.98	-24.11	-94.50	NA	-155.09

based farms is not always positive, especially in the regions of Punjab and Kerala. It is, however, interesting to note from the same table that, when it comes to fully hired bullock labour based farms, the income per unit of land is in most cases negative. Thus, we have no hesitation in drawing the conclusion that, income per unit of land negatively decreasing with more and more hired inputs based tenant operating farms, certainly does not hold true for owner operated farms.

2.3 Attached Labour based Farms

We now turn to results pertaining to permanent farm servants presented in Table 3.

The income per unit of land for all owner operated farms is positively associated with the employment of permanent farm servants along with the employment of hired bullock labour in all States considered. However, in the cases of fully tenant operated farms, the same pattern may not be discerned for different modes of employment of permanent servant and bullock labour.

Now, taking the results of Table 2 and 3 together, we have to draw the following conclusions:

(a) Incomes per unit of land for all types of fully owner operated-casual labour based farms are always lower than the incomes of all types of fully owner operated-permanent farm servant based farms. This is equally true in the cases of all types of fully tenant operated farms. Hence, it follows that employment of permanent farm servant is rather economical for a farmer (owner or tenant) than the employment of casual labour for farm works.

(b) Incomes per unit of land for all types of fully tenant operated-casual labour based farms are mostly non-significant or negative as compared to the incomes of all types of fully tenant operated-permanent farm servant based farms. Thus it follows that decreasing income of tenants results in increasing employment of casual labour and hired bullock labour. If these are substituted by their own family workers or some amount of permanent farm servants as well as their own

TABLE 3

Income per unit of land under purely permanent farm servant based farming for different regions (farm size groups as a whole)

Mode of cultivation	Regions with agl. years						
	Assam 1968-69	Punjab 1969 70	UP 1968-69	AP 1969-70	Tamilnadu 1969-70	Kerala 1962-65	Orissa 1969-70
(1)	(2)	(3)	(4)	(5)	(6)	(7)	(8)
M_9 : Fully permanent farm servant and own bullock labour based owner cultivating farms	840.55	1005.54	2520.88	473.84	1291.71	574.37	998.24
M_{10} : Fully pemanent farm servant and own bullock labour based tenant cultivating farms	232.80	0.82	909.38	64.50	215.88	57.56	325.85
M_{11} : Fully permanent farm servant and fully hired bullock labour based owner cultivating farms	737.19	852.69	2136.88	412.97	1180.93	519.36	767.22
M_{12} : Fully permanent farm servant and fully hired bullock labour based tenant cultivating farms	129.04	—152.03	525.38	3.63	105.10	2.55	94.83

bullock labours, the income of tenants may be greater than the daily income of wage labourer.

3. Concluding Observations

The above analysis of employment of different kinds of labourers relating to the different modes of cultivation brings out the following points:

(a) The incomes of fully owner operated farms will never be negative if any type of reorganisation is made into the system of cultivation, i.e., the cultivation with or without any kinds of hired labour or the cultivation with or without hired bullock labour along with or without any kinds of hired human labour. But the prospect of tenant cultivation lies only with the use of fully family labour and own bullock labour. The use of hired labour and/or hired bullock labour in the tenant operated farms reduce the income of farmer below the level of market wage rate of agricultural labourer.

(b) Among different kinds of hired labour based farming, permanent farm servant based farming may have better prospect into cultivation because of the fact that the incomes of farmers (both owner and tenant) on this mode are higher than that of casual labour based cultivation, and the daily income of permanent farm servant is lower than that of a casual labour.

(c) It may appear from the analysis that the relative importance of different modes of cultivation in agriculture does vary according to the characteristics of the regions. As for example, it is found that in the agriculturally better-endowed regions (Punjab, Kerala etc.) the incomes per unit of land under any type of purely tenant operated farms are non-significant or negative. Family labour income per day in the purely tenant operated farms is lower than than the daily wage rate of agricultural labourer in these two regions. But, the regions which are regarded as relatively less developed, do not bear these patterns.

We may now conclude that the relative importance of different modes of cultivation in agriculture does vary according to the characteristics (both institutional and technological) of the regions. There is enough indication that in the green revolution areas, purely tenant operated farms are not viable. Owner cultivated farms with permanent farm servants may have to prove greater efficiency in terms of employment and income as compared to tenant cultivated farms. But, in the non-green revolution areas tenant cultivated farms have not yet proved ineffective. Thus, it is quite understandable that, with the changing conditions of Indian agriculture virtual elimination of a particular mode and subsequent replacement of another near-by mode takes place. The replacement is made not on a random basis, but follows a particular pattern, as indicated from our analysis. Under this process, the question of transition from 'Feudal[1]' or 'Semi-Feudal'/'Pre-capitalist' to 'Capitalist' mode appears as a natural corollary of the induced changes in the modes of production.

NOTES

1. Special mention may be made of some scholars (e.g., Patnaik 1976; Bardhan, K. 1977; Rudra, 1978) who have extensively discussed this point.

2. We have defined 'modes of production' in agriculture on the basis of the available empirical evidence, provided by various scholars working in this field. Some examples are provided below :

   Thus, Krishna Bharadwaj (1974) wrote: "Such a neat division of the farming system between peasant and capitalist farming, while suggestive, is much too stylised. The link between the predominantly family-based farms and those depending on hired labour exists through the labour market".

   Ashok Rudra (1976), while identifying small farms in his illuminating paper on "Hiring of labour by Poor Peasants", has used the criterion of labour purchase, viz., hiring out of family labour, to define and classify small 'poor peasant farms'. Utsa Patnaik (1976) has used two major criteria—labour hiring and land leasing—to define and classify the different economic classes in agriculture, e.g., in the case of capitalist landlord-labour hiring is greater than rent; in the case of feudal landlord-labour hiring is at most as high as rent; in the case of poor peasants-labour hiring out is greater than rent payment etc.

REFERENCES

Bardhan, Kalpana
(1977) "Rural Employment, Wages and Labour Markets in India: A Survey of Research", *Economic and Political Weekly*, June 25, July 2 & 9.

Bharadwaj Krishna
(1974) *Production Conditions in Indian Agriculture : A Study Based on Farm Management Surveys.* Cambridge University Press, London.

Patnaik, Utsa
(1976) "Class Differentiation within the Peasantry: An Approach to Analysis of Indian Agriculture", *Economic and Political Weekly*, Review of Agriculture, September 25.

Rudra, Ashok
(with M.M. Mukhopadhyay)
(1976) "Hiring of labour by Poor Peasants", *Economic and Political Weekly*, January 10.
(1978) "Class Relations in Indian Agriculture," *Economic and Political Weekly*, June 3, 10 & 17.

Sen, A.K.
(1962) "An Aspect of Indian Agriculture", *Economic Weekly*, Annual Number, February.

Sen, Bhowani
(1962) *Evolution of Agrarian Relations in India.* People's Publishing House, New Delhi.

4

Technology, Croppin Patterng and output

Introduction

There is a general agreement among the Social Scientists of our country that a major breakthrough in agriculture has taken place around 1965 due to the innovation of H.Y.V. Technology. The immediate impact of this kind of technology is expected to change in cropping pattern for increasing output and profit. Thus the farmers will use those crops which are more remunerative as compared to the crops cultivated in the 1950s.

This chapter seeks to examine whether the cropping pattern in agriculture has really changed due to technological innovation. For this purpose we shall undertake a cropwise input-output analysis for some regions of India.

2. Methodology and Data

We have examined the impact of technology upon cropping pattern on the basis of comparisons of Farm Management Data pertaining to different time points in those cases where these are comparable.[1] It may be mentioned that the output per

---

1. In this context see, *Studies in Economics of Farm Management* : Punjab (Ferozepur) 1954-57 & 1967-70; U.P. (Muzaffarnagar) 1954-57 & 1966-69; West Bengal (Hooghly) 1954-57 & 1970-73.

hectare of cultivated area as reported by the F.M.S. has increased during the 1960's over that of the 1950's in the regions referred to above. This increase may be attributed to (a) expansion of acreage, (b) increase in price, and (c) associated changes in the cropping pattern due to the new technology. We have tried to isolate the changes in output due to the two major factors namely "acreage effect" and "price effect". Interaction effect of these two factors upon output has also been worked out. The methodology used for this prupose is as follows.

We have computed for each crop in each region the change in acreage as well as change in price during two time periods for which data are available. Thus, for each crop, the change in output due to acreage alone is given $(\triangle A) \times 0^{50}$ where $\triangle A$ is change in acreage and 0^{50} is the output per hectare in 1950. Similarly, change in output due to price alone is given by $(\triangle 0) \times A^{50}$ where A^{50} is the acreage in 1950 and $\triangle 0$ is the change in per hectare output in value terms. Combining all the crops together we can have the total change in output due to acreage alone and that due to price alone for each region.

We have to take separate account of the output associated with the completely new crops which are grown in the 1970's or in the 1960's but not in the 1950's. We have also to take account of interaction effect of acreage and price upon output by computing $\triangle A \times \triangle 0$ where $\triangle A$ denotes the change in acreage and $\triangle 0$ denotes the change in per hectare output. We now turn to a discussion of the results.

3. Acreage and Price Effects on Farm Output

We shall start by taking a look at the results presented in Tables 1A, 1B, 1C for the region of Punjab. It is seen that except for unirrigated or Desi wheat the acreages of all other crops have increased significantly so that total output due to acreage change alone has increased to a considerable amount.

It is interesting to note that the overall impact of price alone upon output is significantly positive. However, the increased output due to acreage alone is much lower than the increased output due to price alone. The role of new crops with regard to increasing output is also significant which is surely due to the technological developments of the 1960's.

Coming now to the data presented in Tables 2A, 2B and 2C for the region of Uttar Pradesh, it is seen that for some individual crops acreage decreases output, but for all the individual crops price has a positive effect upon increasing output in value terms. Although the new crops contribute towards greater output in the latter period, its effect is lessened by the disappearance of some other crops which were grown in the 1950's. Thus, it appears that price effect is very much pronounced in a positive fashion in this region.

Results of our analysis pertaining to West Bengal (Hooghly) present also the same pattern. It is found from Tables 3A, 3B and 3C that for all the crops, cultivated area has increased so that acreage effect upon output is positive. Price of individual crops during the two time points has also increased so that output has increased considerably. Like other regions, this region is also marked with new crops giving rise to new production, decrease in output due to disappearance of old crops is also present here, but due to lack of data this has not been possible to incorporate. However, at any rate, in the case of West Bengal also the large amount of output seems to have increased over time which is largely due to the changes in price and acreage of new crops.

Thus, the data analysed so far separately for the three regions of India give more or less the same picture of increasing output. That is, price is the most important factor in increasing output in the present days and then comes the role of new crops and acreage expansion of the crops existing in both the time points. Some differentiation however can be made among the regions if we give a close look at the nature of variations of each factor against each region. Thus it is seen that in the case of Punjab, price is the most important factor accounting for change in output; in the case of U.P., increase in price as well as increase in acreage combine to accelerate the changes in output; and in the case of West Bengal, introduction of new crops is the most important factor. If, we think that introduction of new crops and disappearance of old crops due to technology which seems to be increasing output, is the prime condition of changing cropping pattern, then our analysis neither supports nor rejects this proposition at least in the case of U.P. and Punjab where only 3% and 9% of output to the total change in output has taken place in respective two regions. Whereas in the case of West Bengal about 35% of total change in output is explained by the effect of new crops.

We may now turn to examine whether earnings from individual crops can also explain the changes in cropping pattern described above.

4. Output and Cost Effects on Farm Income

Presented in Tables 4A, 4B, 4C are the results of our analysis regarding earnings from individual crops for three regions of India namely Punjab (Ferozepur), U.P. (Muzaffarnagar), and West Bengal (Hooghly) respectively. We have here tried to isolate the changes in income due to the major factors namely "output expansion" and "higher cost of cultivation." Interaction effect of these two factors upon income has also been worked out. The methodology used is the same as in the case of analysing the changes in output presented in Section I.

Thus, change in earning due to output alone is given $(\triangle 0) \times y^{50}$ where y^{50} is the net income per hectare in 1950s and $\triangle 0$ is the change in per hectare output. Similarly, change in earning due to cost effect is calculated as $(\triangle y) \times 0^{50}$ where 0^{50} is the output per hectare in 1950s and $\triangle y$ is the change in per hectare net income.

It is seen from Table 4A that changes in earning is positive in the caae of paddy (new crop), Mexican wheat and American cotton for the region of Punjab. For all other crops in this region, the earning changes negatively due to either higher cost or lower output or both of individual crops. Taking all the crops together, it is seen that change in earning is negative which is largely for the rising costs of cultivation in new crops.

It is therefore interesting to note that the overall impact of cropping pattern upon earning is negative, although for some individual crops its contribution is positive. Thus the so-called new crops which has been possible due to the new kind of technology has not increased income in real sense. It will therefore be fallacious to think that cropping pattern has contributed a lot to the development of agriculture in Punjab.

Coming now to the data presented in Table 4B for the region of U.P., we find that for all the individual crops except old crop (cotton), earning from crop production has increased. This has been possible by the fact that proportionate increase in output is substantially higher than the proportionate increase in cost of cultivation. The role of new crops with respect to total change in earning is also positive. On the basis of these data it may be concluded that increased income results from increase in output with substantial decrease in cost of cultivation. In this case there is an indication that modern farm technology increases the earning of farmers with or without changes in cropping pattern.

West Bengal data which are presented in Table 4C give an interesting picture. Changes in income are positive in the crops existed in both the time points whereas the reverse is found in the case of new crops. It may be recalled from Section 3 that output is positive in the case of new crops. Negative income of new crops thus indicate that relative cost is higher than that of output per unit of land.

5. Effects of Technology and Cropping Pattern on output

Our above analysis suggests that positive changes in output of any crop do not necessarily indicate the prospect of changing cropping pattern in that region. Positive changes in earning are also the prime consideration towards changing cropping pattern. Thus the hypothesis itself suggests that whenever there is a substantial positive increase of both output and earning, it is associated with shifts in cropping pattern and/or new technology.

This hypothesis is confirmed in the cases of West Bengal and Uttar Pradesh where both changes in income and output are positive. In the case of Punjab it lends rather questionable support as change in output is positive but change in income is negative.

The region where the income changes negatively is associated with the fact that the cost of production is positively high than the total value of output. In this case, change in prices of the commodities is found to be positive, but this change is not offset by the additional amount required by the changes in costs of cultivation. In those regions where both income and output are positive are explained either by the propertionate decrease in cost of production or substantial increase in prices. Combined effect of these two factors has led to change in income and output positively.

Coming now to the individual crops, it is observed from the above analysis that new crops always play a positive role in increasing output but not necessarily in increasing net earning. The crops which exist over time such as, irrigated wheat, American cotton, jute etc.. mostly play positive role in increasing output and income. For all other crops e.g., unirrigated wheat, cotton etc., there is a tendency of decrease in income and little increase in output resulting thereby in a questionable prospect of these crops in near future.

Common sense suggests that those crops which are *both* remunerative and productive as far as the soil and climate of the region is concerned will be preferred to other crops (viz.. unirrigated or desi wheat, pulses, cotton etc.) by the farmers. But in practice we observe the co-existence of both remunerative and non-remunerative crops specific to the regions. Possibly same factors other than economic may explain this phenomenon which is beyond our scope in this type of macro analysis.

TABLE 1A

Cultivated area, yield per hectare and price of major crops in Punjab (Ferozepur) : 1955-56 and 1968-79

Crops	*Cultivated area*		*Yield per Hectarc (Qtl.)*		*Price of the commodity per (Qtl.)*	
	1955-56	1968-69	1955-56	1968-69	1955-56	1968-69
(1)	(2)	(3)	(4)	(5)	(6)	(7)
(a) Irrigated or Mexican Wheat	437.40	702.24	10.40	22.46	31.83	77.29
(b) Unirrigated or Desi Wheat	230.91	170.70	6.33	13.63	29.37	69.73
(c) American cotton	251.92	373.44	7.63	8.76	74.87	174.53
(d) Desi cotton	37.07	71.31	6.32	8.07	54.40	124.55
(e) Paddy	–	78.05	—	24.91	—	50.15
(f) Desi maize	—	87.39	—	13.60	—	52.20

TABLE 1 B

Output and cost of production of principal crops in Punjab (Ferozepur) : 1955-56 and 1968 69

(Per Hectare)

Crops	*Output* (*Rs.*)		*Cost of Production*		*Net Income*	
	1955-56	1968-69	1955-56	1968-1969	1955-56	1968-69
(1)	(2)	(3)	(4)	(5)	(6)	(7)
(a) Irrigated or Mexican Wheat	331.05	1736.00	373.92	1618.51	—42.87	117.49
(b) Unirrigated or Desi Wheat	185.92	950.43	174.05	1170.56	11.87	—220.13
(c) American cotton	571.27	1528.85	454.33	1206.02	116.92	322.83
(d) Desi cotton	343.80	1005.09	353.10	1195.01	—9.30	—180.92
(e) Paddy	—	1249.17	—	1164.20	—	84.97
(f) Desi Maize	—	710.28	—	1309.91	—	—599.91

TABLE 2A

Cultivated area, yield per hectare and price of major crops in Uttar Pradesh (Muvaffarnagar) : 1954-55 and 1968-69

Crops	*Cultivated area*		*Yield per hectare (Qtl.)*		*Price of commodities per Qtl.*	
	1954-55	1968-69	1954-55	1968-69	1954-55	1968-69
(1)	(2)	(3)	(4)	(5)	(6)	(7)
(a) Sugarcane planted	110.52	217.92	391.00	515.63	3.60	9.00
(b) Sugarcane ratoon	89.06	140.91	296.00	342.77	3.69	9.00
(c) Unirrigated or Desi Wheat	78.80	350.74	7.06	25.56	38.24	89.01
(d) Irrigated or HYV wheat	165.80	70.21	13.62	34.82	38.55	89.00
(e) Paddy	46.44	130.95	15.00	24.79	26.33	65.07
(f) Maize	71.16	52.83	10.90	13.66	28.90	71.67
(g) Gram	75.14	55.19	7.60	12.09	29.60	90.82
(h) Pea	—	54.35	—	13.33	—	67.44
(i) Gochani	—	22.32	—	18.45	—	80.00
(j) Cotton	12.42	—	4.80	—	63.00	—

TABLE 2B

Output and cost of production of principal crops in Uttar Pradesh (Muzaffarnagar) : 1954-55 and 1968-69

Crops	*Output (Rs.)* 1954-55	1968-69	*Cost of Production* 1954-55	1968-69	*Net Income* 1954-55	1968-69
(1)	(2)	(3)	(4)	(5)	(6)	(7)
(a) Sugarcane planted	1410.00	4641.00	770.00	2194.00	640.00	2447.00
(b) Sugarcane ratoon	1092.50	3085.00	425.00	1217.00	667.50	1868.00
(c) Unirrigated or Desi wheat	270.00	2275.00	287.50	1350.00	—17.50	925.00
(d) Irrigated or HYV wheat	525.00	3099.00	407.50	1474.00	117.50	1625.00
(e) Paddy	395.00	1613.00	252.50	1013.00	142.50	600.00
(f) Maize	315.00	969.00	267.50	903.00	47.50	66.00
(g) Gram	225.00	1098.00	140.00	942.00	85.00	156.00
(h) Pea	—	899.00	—	891.00	—	8.00
(i) Gochani	—	1476.00	—	1294.00	—	192.00
(j) Cotton	302.40	—	202.50	—	100.00	—

TABLE 3A

Cultivated area, yield per hectare and price of major crops in West Bengal Hooghly) 1954-55 and 1970-73

Crops	*Cultivated area*		*Yield per hectare (Qtl)*		*Price of the comm. per Qtl.*	
	1954-55	1970-73	1954-55	1970-73	1954-55	1970-77
(1)	(2)	(3)	(4)	(5)	(6)	(7)
(a) Paddy Aus, Aman or ordinary	91.43	118.57	17.01	19.51	24.00	90.93
(b) Jute	17.69	33.32	14.95	14.53	68.50	176.82
(c) Potato	3.96	23.16	118.17	138.43	NA	49.02
(d) Improved paddy	--	45.02	—	29.07	—	91.52
(e) Wheat	—	8.61	—	15.22	—	98.68
(f) Pulses	9.82	—	6.44	—	NA	—

NA=Not Available

TABLE 3B

Output and cost of production of principal crops in West Bengal (Hooghly) : 1954-55 and 1970-73

(per hectare)

Crops	*Output (Rs.)*		*Cost of production*		*Net income*	
	1954-55	1970-73	1954-55	1970-73	1954-55	1970-73
(1)	(2)	(3)	(4)	(5)	(6)	(7)
(a) Paddy Aus, Aman or ordinary	408.24	1774.04	466.74	1531.54	– 58.50	242.50
(b) Jute	1024.07	2569.18	761.70	1639.14	262.37	930.04
(c) Potato	NA	6785.55	2542.80	6235.17	NA	550.38
(d) Improved paddy	—	2660.74	—	2231.39	—	429.35
(e) Wheat	—	1501.87	—	1649.36	—	—147.49
(f) Pulses	NA	—	185.07	—	NA	—

NA=Not Available

TABLE 1C

Changes in output, Punjab (Ferozepur) between 1955-66 ond 1968-69

Crops	Changes due to old and new crops	Acreage effect $(A^{69}-A^{59})\ 0^{56}$	Price effect $(0^{69}-0^{56})\ A^{56}$	Introduction effect $(A^{69}-A^{56}$ $(0^{69}-0^{56})$	Total changes in output $(A^{69}\text{-}0^{69}\text{-}(A^{56}x0^{56}$
(1)	(2)	(3)	(4)	(5)	(6)
New crops					
(a) Paddy	97497.72	—	—	–	97497.79
(b) Desi maize	62071.37	—	—	—	62071.37
Present both in 1950's and 1970's					
(a) Irrigated or Mexican wheat		87675.28 (8.16)	614525.13 (57.20)	372086.95 (34.63)	1074287.40 (100.00)
(b) Unirrigated or Desi wheat		—11194.24 (—9.38)	176533.00 (148.00)	—46031.15 (—38.58)	119307.61 (100.00)
(c) American cotton		69420.73 (16.25)	241233.55 (56.50)	116365.12 (27.25)	427019.41 (100.00)
(d) Desi Cotton		11771.71 (19.97)	24514.02 (41.60)	22642.57 (38.42)	58928.31 (100.00
Total	159569.09 (8.67)	157673.48 (5.67)	1056805.60 (57.40)	465063.49 (25.28)	1839111.80 (100.00)

Note : Figures in bracket indicate percentage of total change in output.

TABLE 2 C

Changes in output, Uttar *Pradesh* (Muzaffarnagar) between 1954-55 and 1968-69

Crops	Changes due to old and new crops	Acreage effect $(A^{69}-A^{55})\ 0^{55}$	Price effect $(0^{69}-0^{55})\ A^{55}$	Introduction effect $(A^{69}-A^{55})\ (0^{69}-0^{55})$	Total change in output $(A^{69}-0^{69}-(A^{55} x\ 0^{55})$
(1)	(2)	(3)	(4)	(5)	(6)
New crops					
a) Pea	48860.65	—	—	—	48860.65
b) Gochani	32944.32	—	—	—	32944.32
Old crops					
a) Cotton	3758.83	—	—	—	3758,83
Present both in 1950's and 1970's					
a) Sugarcane planted		151434.00 (17.70)	357090.12 (41.73)	347009.40 (40.56)	855533.50 (100.00)
b) Sugarcane ratoon		56646.12) (16.78)	177452.05 (52.60)	103311.12 (30.61)	337409.30 (100.00)

contd......

1	2	3	4	5	6
c)	Unirrigated or Desi wheat	– 73423.80 (9.45)	157994.00 (20.34)	545239.70 (70.20)	776657.50 (100.00)
d)	Irrigated or H.Y.V. wheat	– 50184.75 (38.44)	426769.20 (326.93)	—246048.66 (—188.49)	130535.79 (100.00)
e)	Paddy	33381.45 (17.30)	56563.92 (29.32)	102933.18 (53.36)	192878.55 (100.00)
f)	Maize	—5773.95 (20.06)	46538.64 (161.72)	—11987.82 (41.65)	28776.87 (100.00)
g)	Gram	—4488.75 (10.27)	65597.22 (150.13)	—17416.36 (—39.86)	43692.12 (100.00)
Total	78046.14 (3.19)	254437.92 (10.41)	1288005.00 (52.71)	823040.60 (33.68)	2443529.50 (100'00)

Note : Figures in parenthesis indicate percentage of total change in output.

TABLE 3C

Change in output, West Bengal (Hooghly) between 1954-55 and 1970-73

Crops	Change due to old and new crops	Acreage effect $(A^{73}-A^{55})\, 0^{55}$	Price effect $(0^{73}-0^{55})\, A^{55}$	Interaction effect $(A^{73}-A^{55})\,(0^{73}-0^{55})$ $Ax^{73}-0^{73})$	Total changes in output $(Ax^{55}-0^{55})$
1	2	3	4	5	6
New crops					
a) Improved paddy	119786.51	—	—	—	119786.51
b) Wheat	12931.10	—	—	—	12931.00
Old crops					
a) Pulses	NA				NA
Present both in 1950's and 1970's					
a) Paddy Aus, Aman ordinary		11079.63 (6.40)	124875.09 (72.17)	37067.81 (21.42)	173022.54 (100.00)
b) Jute		16518.25 (24.01)	27333.00 (39.74)	24922.62 (36.23)	68773.87 (100.00)
c) Potato		NA	NA	NA	NA
Total	132717.61 (35.43)	27597.88 (7.36)	152208.09 (40.64)	61990.43 (16.55)	374514.02 (100.00)

Note : Figures in parenthesis indicate percentage of total change in output.

NA = Not Available.

TABLE 4 A

Changes in earning, Punjab (Ferozepur) betweeu 1955-56 and 1968-69

Crops	Changes due to old and new crops	Output effect $(0^{69}-0^{56})\ y^{55}$	Cost effect $(y^{69}-y^{56})\ 0^{56}$	Interaction effect $(0^{69}-0^{56})-(y^{69}-y^{56})$	Total change in earning $(0^{69}xy^{69})-(0^{56}xy^{56})$
1	2	3	4	5	6
New crops					
a) Paddy	106141.97				106141.97
b) Desi maize	--426104.07				—426104.07
Present both in 1950's and 1970's					
a) Irrigated or Mexican wheat		--60230.20 (−27.60)	53087.71 (24.33)	225297.78 (103.27)	218154.75 (100.00)
b) Unirrigated or Desi wheat		9074.73 (4.30)	—43133.44 (20.40)	− 177366.32 (83.89)	—211425.02 (100.00)
c) American cotton		111960.25 (26.24)	117630.20 (27.56)	197175.29 (46.20)	426765.75 (100.00)
d)Desi cotton		—6149.99 (3.27)	—62097.16 (33.08)	—119442.20 (63.63)	—187689.35 (100.00)
Total	—319962.10 (—431.46)	54654.79 (73.70)	65486.77 (88.30)	125664.55 (169.46)	--74155.87 (100.00)

Note : Figures in parenthesis indicate percentage of total change in earning.

TABLE 4B

Changes in earning, Uttar Pradesh (Muzaffarnagar) between 1954-55

crops	Changes due to old and new crops	Output effect $(0^{69}\text{-}0^{55})\ y^{55}$	Cost effect $(y^{69}\text{-}y^{55})\ 0^{55}$	Interaction effect $(0^{69}\text{-}0^{55})\text{-}(y^{69}\text{-}y^{55})$	Total change in earning $(0^{69}xy^{69})\text{-}(0^{55}xy^{55})$
1	2	3	4	5	6
New crops					
a) Pea	7192.00				7192.00
b) Gochani	283392.00				283392.00
Old crop					
a) Cotton	—30240.00				O30240.00
Present both in 1950's and 70's					
1	2	3	4	5	6
a) Sugarcane planted		2067840.00 (19.78)	2547870.00 (24.37)	5838417.00 (55.84)	10454127.00 (100.00)
b) Sugarcane ratoon		1329993.70 (26.42)	1311546.20 (26.05)	2391996.20 (47.52)	5033536.30 (100.00)

contd......

1	2	3	4	5	6
c)	Unirrigated or Desi wheat	—35087.50 (—1.66)	254475.00 (12.06)	1889712.50 (89.59)	2109100.00 (100.00)
d)	Paddy	173565.00 (19.04)	180712.50 (19.82)	557235.00 (61.13)	911512.50 (100.00)
e)	Irrigated or HYV wheat	302445.00 (6.08)	791437.50 (15.91)	3880305.00 (78.00)	4974187.50 (100.00)
f)	Maize	31065.000 (63.40)	5827.50 (11.89)	12099.00 (24.70)	48991.50 (100.00)
g)	Gram	74205.00 (48.76)	15975.00 (10.50)	6183.00 (40.73)	152163.00 (100.00)
Total	320824.00 (1.33)	3944026.20 (16.43)	510784.70 (21.27)	14631746.70 (60.95)	24004441.30 (100.00)

Note : Figures in parenthesis indicate percentage of total change in earning.

TABLE 4C

Changes in earning, West Bengal (Hooghly) between 1954-55 and 1970-73

Crops	Changes due to old and new crops	Output effect $(0^{73}-0^{55})\ Y^{55}$	Cost effect $(0^{73}-0^{55})\ 0^{55}$	Interaction effect $(0^{73}-0^{55})\ (Y^{73}-Y^{55})$	Total change in earning $(0x^{73}y^{73})-(0x^{55}y^{55})$
1	2	3		5	6
New crops					
a) Wheat	—221510.80	—	—	—	—221510.80
Old crops					
Pulses	NA				NA
Present both in 1950's and 1970's					
a) Paddy Aus, Aman or ordinary —		—79899.30 (−17.60)	122880.24 (27.06)	41110.580 (90.53)	454086.74 (100.00)
b) Jute		405390.51 (19.11)	683740.81 (32.24)	1031623.50 (48.64)	2120754.90 (100.00)
c) Potato		NA	NA	NA	NA
Total	− 221510.80 (—9.41)	325491.21 (13.84)	806621.05 (34.27)	1442729.30 (61.30)	2353330.84 (100.00)

Note : Figures in parenthesis indicate percentage of total change in earning.

NA = Not Available.

5

Size of Holding, Farm Profit and Labour Exploitation

1. Introduction

This chapter seeks to present some quantitative results pertaining to the inter-relations between (i) the composition of cost of cultivation in terms of labour cost and material cost; (ii) the intensity of labour cost to cultivation; (iii) the productivity of labour and the wage rate; (iv) the productivity of land and the cost of cultivation; and (v) the intensity of profit according to the size of holding. With the help of these quantitative results we have proposed to examine the validity of certain hypotheses which have generated a great deal of discussions among the theoretical economists. The hypotheses are listed below.

Hypothesis 1 (H_1) : Larger farms have a higher proportion of profit.

Hypothesis 2 (H_2) : As farm size increases, productivity of land decreases but the rate of profit increases by minimising the costs of cultivation.

Hypothesis 3 (H_3) : As farm size increases, average productivity of labour increases but the wage rate decreases.

Hypothesis 4 (H_4) : Labour is the "soft input" for acquiring more profit.

2. Data and Methodology

The validity of the hypotheses have been examined here with the help of three sets of Farm Management data. Data include three regions of India, namely Punjab (Ferozepur) for the year 1968-69, West Bengal (Hooghly) for the years 1970-73, and Assam (Nowgong) for the year 1968-69.

For the sake of analytical convenience the different cost items for cultivation (cost A_1, cost A_2, cost B, and cost C as defined by Farm Management Survey Reports) have redefined as follows :

Total cost=Cost A+Cost B+Cost C

Cost A=Cost of family labour+cost of hired labour

Cost B=Cost of material inputs including that of bullocks, manures and fertilisers, seeds, depreciation and interest on working capital

Cost C= Land revenue, rent paid for land taken on lease, rental value of own land, interest on own fixed capital

Net income (profit/loss)=Gross income—total input cost

Since farm profit or loss is determined by the surplus or deficit of output over costs including the imputed value of labour, the different types of profit according to the influences of different types of costs can be calculated. These can be stated as follows :

Profit AB= Total output—(Cost A+Cost B) (assuming cost C remains zero)

Profit BC= Total output—(Cost B+Cost C) (assuming cost A remains zero)

Profit AC= Total output—(Cost A+Cost C) (assuming cost B remains zero).

Thus, if we exclude the contribution of labour, that is cost A from the total output, i.e., when cost A is made zero for all farms, the profit essentially gives us the contributions of cost B and cost C. Similarly for the other cost items. The above have been calculated on the basis of per hectare cultivated area.

Now, in order to determine the exploitation rate corresponding to each farm, we can use the following formula :

$$\text{Rate of exploitation} = \frac{\text{Total surplus value}}{\text{Total surplus value excluding the labour value}}$$

This is invariant under change of scale of measurement.

3. Results of Analysis

We shall put our findings with regard to the different hypothese separately for the three different regions. As they refer to certain selected areas and certain selected time points, the results do not *claim any definitive refutation or confirmation* of the hypotheses for the country as a whole. Nevertheless, hopefully, the insights obtained through the analysis will provide for sufficient stimulation in this area of research.

3.1 Punjab

Presented in Tables 1.1 to 1.6 are the results of analysis for the region of Punjab (Ferozepur).

TABLE 1.1

Profit/Loss by Farm Size (Punjab) 1968-69

Farm size group (hectare)	Output per hectare (Rs.)	Cost per hectare (Rs.)	Profit/ Loss	Share of Profit
(1)	(2)	(3)	(4)	(5)
I Below 6	1708.60	1736.84	—28.24	—4.51
II 6—9	1631.99	1564.84	67.15	10.72
III 9—14	1556.92	1625.75	—68.83	—10.98
IV 14—24	1573.34	1428.60	144.74	23.10
V 24 & Above	1958.17	1446.39	511.78	81.67
All sizes	1687.06	1596.22	90.84	100.00

Table 1.1 has been prepared in order to examine the relationship between size of farms and farm profit. It is seen from Table 1.1 that *profit per* hectare increases *as farm size goes up*. Output per hectare also goes up with farm sizes, but the reverse is true in the case of total cost of cultivation per hectare. The cost data show that the larger sized farms put in less amount of cost compared to the smaller sized farms, but owing to higher productivity of land, the rate of profit as well as share of profit seem to be higher in the former farms than the latter. This is also confirmed from the correlation matrix (Table 1.2) indicating that the correlation between profit (Y) and average size of holding (X_1) is positively high, but the correlations between size of holding (X_1) and various cost items are negatively high (except cost C). Thus, both the hypotheses H_1 and H_2 are accepted for the region of Punjab.

TABLE 1.2

Correlation Coefficient Matrix : Punjab 1968-69

	Y	X_1	X_2	X_3	X_4
Y	1.00000	0.93926	—0.95068	—0.74687	0.66248
X_1	0.93926	1.00000	—0.94888	--0.83040	0.69175
X_2	—0.95068	—0.94888	1.00000	0.91398	—0.54861
X_3	—0.74687	—0.83040	0.91398	—1.00000	—0.27476
X_4	—0.66248	0.69175	—0.54861	—0.27476	1.00000

(Y=Profit, X_1=Average holding, X_2=Cost A, X_3=Cost B, X_4=Cost C)

An interesting finding that we get from the analysis of various cost data in relation to farm sizes is that larger sized farms spend less in Cost A (ie, Cost of human labour employment) and more in Cost C (ie, Costs on fixed capital) whereas smaller sized farms do the reverse (Table 1.3)

TABLE 1.3

Distribution of Each by Farm Size Groups : Punjab, 1968-69

Farm size group (hectare)	Per hectare			
	Cost A (Rs.)	Cost B (Rs.)	Cost C (Rs.)	Total (Rs.)
(1)	(2)	(3)	(4)	(5)
I Below 6	468.45 (24.47)	671.22 (23.53)	591.17 (19.64)	1736.84 (22.26)
II 6—9	394.70 (20.62)	586.01 (20.36)	584.13 (19.41)	1564.84 (20.05)
III 9—14	426.53 (22.28)	589.12 (20.46)	610.10 (20.27)	1625.75 (20.84)
IV 14—24	355.76 (18.59)	520.94 (18.10)	551.90 (18.34)	1428.60 (18.31)
V 24 & above	268.78 (14.04)	505.21 (17.55)	672.40 (22.34)	1446.39 (18.54)
Total	1914.22 (100.00)	2878.50 (100.00)	3009.70 (100.00)	7802.42 (100.00)

Thus, it follows from the data that larger size farms maximise profit by minimising the Cost of human labour in particular.

Coming now to the hypothesis H_3, it is seen from Table 1.4 that there is a sharp fall in the average number of labourers employed per hectare and also the average labour day per hectare in the larger sized farms. However, there is an Indication that both average productivity of labour as well as daily wage rate of labourers are higher in the larger sized farms than the smaller ones. It indicates that though the larger size farms minimise the labour cost, they simultaneously optimise output with the help of technology and hence are able to maximise profit. Thus, hypothesis H_3 fails to receive confirmation at least in the case of Punjab.

TABLE 1.4

Wage Rate and Productivity by farm size : Punjab 1968-69

Farm size group (hectare)	Av. size of holding (hectare)	Av. no. of labour employed per (hect.)	Av. wage rate per day per labourer	Av. labour productivity	Labour days per hectare
(1)	(2)	(3)	(4)	(4)	(6)
I below 6	4.37	66.15	4.38	18.75	91.10
II 6—9	7.29	61.13	4.57	20.69	78.86
III 9—14	11.29	67.25	5.19	18.13	85.86
IV 14—24	17.66	52.59	5.27	23.55	66.80
V 24 &above	35.34	45.72	5.36	39.34	49.78
All sizess	12.31	59.73	4.90	21.79	77.41

$$\text{Wage rate} = \frac{\text{Hired labour cost}}{\text{Hired labour cost}}$$

TABLE 1.5

Profit per hectare according to different cost concept : Punjab 1968-69

Farm size (hectare)	Total profit (Rs.)	Profit AB (Rs.)	Profit AC (Rs.)	Profit BC (Rs.)
(1)	(2)	(3)	(4)	(5)
I Below 6	—28.24	562.93	648.98	440.21
II 6—9	67.15	651.28	653.16	461.85
III 9—14	—68.83	541.27	520.29	357.70
IV 14—24	144.74	696.64	665.68	500.50
V 24 & above	511.78	1184.18	1016.99	780.56
All sizes	90.84	698.87	682.07	487.80

Since farm profit is determined by the surplus and different inputs or different types of costs are directly related to it, it will be interesting to examine the types of relationships between various cost items and farm profits. In this context, we find from data presented in Table 1.5 that profit per hectare of larger sized farms does not decrease when labour cost remains zero. In fact, data show that the larger farms make always higher profit compared to the smaller sized farms irrespective of any cost items remaining zero. This becomes possible owing to the relatively higher productivity of land in the larger sized farms. Smaller sized farms, on the other hand, get less profit owing to the high expenditure on cost A (labour cost) over the output per unit of land. Thus the labour input cannot be considered a 'soft input' for acquiring profit particularly by the larger sized farms in the case of Punjab. Here, technology or more intense use of material inputs may be considered as the soft input for the purpose of farm profitability.

TABLE—1.6

Rate of increment of profit owning to the exclusion of different costs : Punjab, 1968-69

Farm size (hectare)	Total profit	Profit AB Total profit	Profit AC Total profit	Profit BC Total profit	Total Profit BC (Rate of expln.	Index
(1)	(2)	(3)	(4)	(5)	(6)	(7)
I Below 6	—20.24	—19.934	—22.981	—15.588	—0.064	0
II 6—9	67.15	9.699	9.727	6.878	0.145	0.209
III 9—14	—68.83	—7.864	—7.559	—5.197	—0.193	—0.128
IV 14—24	144.74	4.813	4.599	3.458	0.289	0.353
V 24 & above	511.78	2.314	1.987	1.525	0.656	0.720
All sizes	90.84	7.693	7.508	5.369	0.186	0.250

Table 1.6 has been prepared in order to compare the exploitation rates over various size classes. The method of computation has been discussed in the previous section. Table 1.6 shows that for all practical purposes, the smallest size group of farms has been taken as the norm, i.e., the exploitation rate in this size group is assumed to be zero. On the basis of this assumption, we find from our analysis that the rate of exploitation is significantly higher in the larger sized farms than the other size classes. It means that the larger sized farms have decisive advantages over the smaller ones in respect of their own capital as well as credit facilities, resulting thereby to acquire relatively higher profit by exploiting efficiently the productive forces.

TABLE 2.1

Profit/Loss by Farm Size : West Bengal (Hooghly) 1970-73

Farm size group (hectare)	Output per hectare (Rs.)	Cost per hectare (Rs.)	Profit/ Loss (Rs.)	Share of profit
(1)	(2)	(3)	(4)	(5)
I 0.01—0.50	3332.11	3160.06	172.05	5.72
II 0.51—1.00	3378.82	2930.89	447.93	14.90
III 1.01—1.50	3565.34	3006.61	558.73	18.58
IV 1.51—2.00	3100.41	2688.75	411.66	13.69
V 2.01—8.00	3075.33	2760.24	315.09	10.48
VI 8.01—4.00	4103.67	3739.56	364.11	12.11
VII 4.01—6.00	2896.61	2802.00	94.61	3.15
VIII 6.01 &	3070.90	2428.45	642.45	21.37
All sizes	3334.90	2929.68	405.22	100.00

3.2 West Bengal

In this section we propose to present some data about the variations of farm profits according to the various farm size groups for a region of West Bengal. Data presented in Table 2.1 shows that there is no particular pattern of rate of profit being higher in the larger sized farms. In fact, apart from the biggest sized farms, there is a tendency of farm profitability being higher in the smaller sized classes also. Data further show that both the output and cost per hectare in the smaller sized farms are relatively higher than the larger sized farms. This indicates that the larger sized holdings do not necessarily take initiative to maximise profit through maximisation of output and minimisation of farm costs.

TABLE 2.2

Distribution of Each Cost by Farm Size Group : West Bengal (Hooghly) 1970-73

Farm size	Per Hectare			
group (hectare)	Cost A (Rs.)	Cost B (Rs.)	Cost C (Rs.)	Total (Rs.)
(1)	(2)	(3)	(4)	(5)
I 0.01—0.50	778.45 (15.44)	984.58 (13.17)	1397.03 (12.71)	3160.00 (13.44)
II 0.51—1.00	645.90 (12.81)	916.95 (12.26)	1368.04 (12.44)	2930.89 [12.46)
III 1.01—1.50	626.07 (12.41)	980.55 (13.11)	1399.99 (12.73)	3006.61 12.79)
IV 1.51—2.00	581.40 (11.53)	877.57 (11.73)	1229.78 (11.18)	2688.75 (11.43)
V 2.01—3.00	526.70 (10.44)	855.38 (11.44)	1378.16 (12.53)	2760.24 (11.74)
VI 3.01–4.00	745.41 (14.78)	1443.18 (19.30)	1550.97 (14.11)	3739.56 (15.90)
VII 4.01—6.00	726.50 (14.41)	587.52 (7.76)	1487.98 (13.53)	2802.00 (11.91)
VIII 6.01&above	412.30 (8.18)	832.46 (11.13)	1183.69 (10.77)	2428.45 (10.33)
Total	5042.73 (100.00)	7478.19 (100.00)	1099.64 (100.00)	2311.56 (100.00)

The above odservations are confirmed by the results of correlation matiix (Table 2.3). Table 2.3 shows that the correlation between profit (Y) and average size of holding (Xi) is very low. The correlations between size of holding and various cost items are in most cases negatively low. Thus the results of our analysis for the region of West Bengal do not suggest either refutation or confirmation of the hypotheses H_1 and H_2 presented above.

TABLE 2.3

Correlation Coefficient Matrix

	Y	X_1	X_2	X_3	X_4
Y	1.00000	0.16261	—0.71000	0.20789	—0.58341
X_1	0.16261	1.00000	—0.45371	—0.23688	—0.15646
X_2	—0.71000	—0.45371	1.00000	0.31675	0.79502
X_3	0.20789	- 0.23688	0.31675	1.00000	0.39408
X_4	—0.58341	—0.15646	0.79502	0.39408	1.00000

(Y=Profit, X_1=average holding, X_2=Cost A, X_3= Cost B, X_4=Cost C)

TABLE 2.4

Wage Rate and Productivity by Farm Size : West Bengal (Hooghly) 1970-73

Farm size group (hectare)	Av. size of holding (hectare)	Av. no. of. labour employed per hectare	Av. wage rate per day per labourer	Av. labour productivity	Labour days per hectare
(1)	(2)	(3)	(4)	(5)	(6)
I 0.01—0.50	0.29	3.57	8.19	38.91	85.64
II 0.51—1.00	0.78	1.79	8.26	46.37	72.87
III 1.01—1.50	1.23	1.46	7.44	43.90	81.21
IV 1.51—2.00	1.72	1.29	10.92	42.37	73.18
V 2.01—3.00	2.48	0.95	8.88	49.39	62.27
VI 3.01—4.00	3.48	0.86	12.38	48.97	83.79
VII 4.01—6.00	4.98	0.90	19.27	58 26	49.72
VIII 6.01&above	6.63	0.68	13.42	57.95	52.99
All sizes	1.21	1.37	9.83	46.32	71.99

Coming now to the hypothesis H_3, we find from Table 2.4 that as farm size increases average productivity of labour also increases and wage rate per labourer per day seems to indicate a tendency to increase with the size of holdings. However, intensity of labour employment per unit of cultivated area declines with the increase of size of holdings. Thus, West Bengal data do not support unrestrictedly the hypothesis H_3.

Now to the question as to whether the farms acquire profit by utilising the underpaid services of labourers, we find from Table 2.5 that such relationship cannot be drawn in the case of West Bengal. It is seen that there is no systematic pattern of profit rates increasing if the labour cost is totally excluded from output (i.e., profit BC). This is also true in cases of other types of profit (i.e., profit AB or profit AC). It is clear from these data that the hypothesis H_4 cannot be accepted at least in the case of West Bengal.

TABLE 2.5

Profit per hectare according to different cost concept : West Bengal (Hooghly) 1970-73

Farm size group (hectare)	Total Profit (Rs.)	Profit AB (Rs.)	Profit AC (Rs.)	Profit BC (Rs.)
(1)	(2)	(3)	(4)	(5)
I 0.01—0.50	172.05	1569.08	1156.63	950.50
II 0.51—1.00	447.93	1815.97	1364.88	1093.83
III 1.01—1.50	558.73	1958.72	1539.28	1184.80
IV 1.51—2.00	411.66	1641.44	1289.23	993.06
V 2.01—3.00	315.09	1693.25	1170.47	841.79
VI 3.01—4.00	364.11	1915.08	1807.29	1109.52
VII 4.01—6.00	94.61	1582.59	682.13	821.11
VIII 6.01 & above	642.45	1826.14	1474.91	1054.75
All sizes	405.22	1752.62	1379.24	1013.48

Furthermore, the results of analysis suggest that the farms (not necessarily the bigger sized farms) which acquire higher rates of profit are guided by the principle of increasing productivity of land and simultaneously decreasing in cost of cultivation particularly the labour cost. This is clear from the data of Table 2.6 which show that the rate of labour exploitation is of course higher in the larger sized farms, but the rates are equally high in the smaller size groups.

TABLE 2.6

Rates of increment of profit owing to the conclusion of different costs

Farm size group (hectare)	Total profit / Total profit	Total profit / Profit AB	Total profit / Profit AB	Total profit / Profit BC	Profit / Profit BC (rate of exploitation)	Index
(1)	(2)	(3)	(4)	(5)	(6)	(7)
I 0.01—0 50	172.05	9.12	6.72	5.52	0.18	0
II 0.51—1.00	447.93	4.05	3.05	2.44	0.41	0.23
III 1.01—1.50	558.73	3.51	2.75	2.12	0.47	0.29
IV 1.51—2.00	411.66	3.99	3.13	2.41	0.41	0.23
V 2.01 – 3.00	315.09	5.37	3.71	2.67	0.37	0.19
VI 3.01—4.00	364.11	5.76	4.96	3.05	0.33	0.15
VII 4.01—6.00	94.61	16.73	7.21	8.68	0.11	0.07
VIII 6.01 & Above	642.45	2.84	2.29	1.64	0.61	0.43
All Sizes	405.22	4.32	3.40	2.50	0.40	0.22

Thus, the results of analysis on the basis of West Bengal sample give somewhat different picture which at least do not support the neo-classical idea that as farm size increases, labour cost gets increased but, yield per hectare and output per labour diminishes, the net result would be increasing quantities of farm profitability.

Assam : We have presented our results of analysis in Tables 3.1 to 3.6 for region of Assam.

TABLE 3.1

Profit/loss by farm size : Assam (Nowgong), 1968-69

Farm size group (hectare)	Output per hectare (Rs).	Cost per hectare (Rs)	Profit/loss (Rs)	Share of profit (Rs)
1	2	3	4	5
0.01—1.82	1543.97	1492.24	51.73	10.52
1.83–2.43	1370.47	1237.36	133.11	27.08
2.44—3.24	1306.68	1238.50	68.18	13.87
3.25–4.45	1175.79	1017.81	157.98	32.14
4.46& above	1057.19	976.60	80.57	16.39
All sizes	1215.53	1114.76	100.77	100.00

In relation to first hypothesis, as Table 3.1 shows, profit is not directly proportional to the farm size in the case of Assam. There is a trend of the smaller size groups acquiring more profit than the largest size group. This shows that the big size of the holding does not necessarily mean that the farm will be able to minimise costs with maximisation of output and the profit will be more. This may be owing to the fact that there may be some optimal holdings where the minimisation of cost with maximisation of output can make possible a maximum level of output. What may happen sometimes is that the big holdings are unmanageable and, therefore, optimisation is not possible there.

Thus our data do not lend support to the first hypothesis in the case of Nowgong, Assam. This is also confirmed from the correlation matrix (Table 3.3) showing that the correlation between profit (y) and average size of holding (X_1) is very low.

Now let us examine the second hypothesis which examines whether it is possible to earn more profit whithout optimising output. That is, we examine essentially for which factors the total cost is minimised in connection with maximum profit. For this purpose, let us analyse the nature of the relations among Cost A, Cost B, and C presented in Table 3.2.

It is seen from Table 3.2 that farms that are making more profit actually have got the tendency to spend less in Cost A, than in Cost C and lastly in Cost B compared with others. To spend less in Cost C is not new phenomenon, because big farmers usually do not take lease land from others. Therefore, they spend less on items such as rent and revenue. But what is new, perhaps, is that as farm size increases, labour cost decreases very sharply, much more than material cost, as will be found from the same table.

TABLE 3.2

Distribution of each cost by farm size group : Assam (Nowgong) 1968-69

Per hectare

Farm size group (hectare)	Cost A (Rs.)	Cost B (Rs.)	Cost C (Rs.)	Total (Rs.)
(1)	(2)	(3)	4	(5)
I 0.01—1.82	417.98 (24.45)	278.62 (24.51)	795.64 (25.53)	1492.24 (25.03)
II 1.83—2.43	351.78 (20.57)	240.94 (21.20)	644.64 (20.69)	1237.36 (20.75)
III 2.44—3.24	367.24 (21.48)	242.13 (21.30)	629.13 (20.19)	1238.50 (20.77)
IV 3.25—4.45	292.54 (17.11)	193.90 (17.06)	531.37 (17.05)	1017.81 (17.07)
V 6.46 & Above	280.32 (16.39)	181.01 (15.93)	515.27 (16.54)	976.60 (16.38)
Total	1709.86 (100.00)	1136.60 (100.00)	3116.05 (100.00)	5962.51 (100.00)

TABLE 3.3

Correlation Coefficient Matrix

	Y	X_1	X_2	X_3	X_4
Y	1.00000	0.11729	—0.11729	—0.56608	—0.54072
X_1	0.11729	1.00000	—0.86540	—0.89774	—0.83541
X_2	—0.56608	—0.86540	1.00000	0.99465	0.97416
X_3	—0.48972	—0.89774	0.99465	1.00000	0.97229
X_4	—0.54072	—0.83541	0.97416	0.97229	1.00000

(Y=profit, X_1=average holding, X_2=Cost A, X_3=Cost B, X_4=Cost C)

Further, the results of Table 3.3 confirm the following :

(1) As farm size increases, profit from the farm does not necessarily increase, because the correlation between farm size and profit is very small.

(2) As labour costs decrease, profit increases as the correlation coefficient is negative but high.

(3) As profit increases, material costs on the farm increase compared with labour costs. The correlation coefficient is positive but low.

Thus, our data do not support the second hypothesis also. However, at the same time the data confirm the fourth hypothesis, that is, labour input plays a dominant role in increasing profit. Interestingly, the relation between farm size and profit shows that we cannot identify the big farmers or high profit earning groups in terms of the possession as land. This is also established in Table 3.4.

It is seen from Table 3.4 that there is a sharp fall in the average number of labourers employed per hectare and also the average labour day per hectare and wage rate per day per labourer on big holdings. Though the biggest size farms minimised the labour cost it could not optimise output and hence failed to maximise profit.

TABLE 3.4

Wage rate and productivity by farm size : Assam (Nowgong) 1968-69

Farm size group (hectare)	Average size of holding	Av. no. of labour employed per hect.	Av. wage rate per day per labourer	Av. labour productivity	Labour days per hectare
(1)	(2)	(3)	(4)	(5)	(6)
I 0.01—1.82	1.35	1.56	1.86	12.78	120.85
II 1.83—2.43	2.18	1.08	2.10	13.59	100.83
III 2.44—3.24	2.78	1 05	1.44	12.46	104.86
IV 3.25—4.45	3.77	0.82	1.82	14.11	83.33
V 4.46 & above	6 37	0.57	1.30	13.14	80.43
All sizes	3.22	0.88	1.68	13.20	72.06

We can draw the following conclusions on the basis of the data presented in Table 3.4.

(1) As farm size increases, average productivity of labour increases but the wage rate decreases, which obviously confirms the third hypothesis.

(2) Profit is largely determined by minimising labour costs and not by the maximisation of output, which confirms the second and fourth hypotheses.

Let us now try to examine how curtailing the labour input affects farm profit. If we exclude the contribution of labour, that is Cost A, from the total output, that is when Cost A is made zero for all farms (the profit essentially gives us the contributions of Cost B and Cost C), we find a steady fall in profit with an increase in farm size (Table 3.5). We have already seen that a big farm may not always be able to create a significant amount of profit. Here we find that a big farm gets little profit if it excludes the labour cost. The small farm does not get more profit when Cost B, that is, material cost, is made zero, and certainly it can get much higher profit if costs of land rent, revenue and so on (Cost C) are excluded from the total output.

Thus from the above observations it is clear that the small farm gets less profit owing to the high expenditure on Cost C and Cost A and the remaining farms gets more profit compared with small farms owing to the minimisation of Cost A and Cost C, where Cost B remains more or less constant for all the size groups.

TABLE 3.5

Profit per hectare according to different cost concepts

Farm size (hectare)	Total Profit	Profit AB (Rs)	Profit AC (Rs)	Profit BC (Rs)
(1)	(2)	(3)	(4)	(5)
I 0.01—1.82	51.73	847.87	430.35	469.71
II 1.83—2.43	133.11	777.75	277.05	484.89
III 2.44—3.24	68.18	697.31	310.31	435.42
IV 3.25—4.45	157.98	689.35	351.88	450.52
V 4.46 & Above	80.59	595.36	261.60	360.91
All sizes	100.77	682.49	312.75	421.83

Now, in order to compare the exploitation rate, an index has been calculated in Table 3.6. It shows that for all practical purposes the smallest size group of farms has been taken as the norm, that is, the exploitation rate in this size group is zero. From this the ranking of farm size groups in respect of the rate of exploitation of labour can be done as follows : 1. size group IV, 2. size group II, 3 size group V, 4. size group III. Thus, the hypothesis H_4, once again, is not compatible in the case of larger sized farms but this gets confirmed in those size-groups which acquire proportionably higher farm profit.

TABLE 3.6

Rates of increment of profit owing to exclusion of different costs

Farm size (hectare)	Total profit	Profit AB profit	ProfitAC profit	Profit BC profit	Profit profit BC (Rate of exploitn)	Index
(1)	(2)	(3)	(4)	(5)	(6)	(7)
I 0.01—1.82	51.73	16.381	6.386	9.080	.11086	0
II 1.83—2.43	133.11	5.843	2.810	3.643	.27449	.16436
III 2.44—3.25	68.18	10.227	4.551	6.386	.15659	--04649
IV 3.25—4.45	157.98	4.364	2.227	2.852	.35063	.24050
V 4.46&Above	80.59	7.394	3.246	4.478	.22331	.11318
All sizes	100.57	6.773	3.114	4.186	.23889	.12876

CONCLUSION

The conclusions with respect to H_1, H_2, H_3 and H_4 for the three regions under study are presented in Tabular from in Table 4.

TABLE 4

Summary of the Results of Analysis

Hypotheses	Punjab (Ferozepur): 1968-69	West Bengal (Hooghly) 1970-73	Assam (Nowgong) 1968-69
(1)	(2)	(3)	(4)
H_1: Larger farms have a higher proportion of profit	Accepted	Undecided (no distinct pattern)	Rejected

(1)	(2)	(3)	(4)
H_2: As farm size increases, the rate of profit increases by minimisation of cost and maximisation of output	Accepted	Undecided (no distinct pattern)	Rejected
H_3: As farm size increases, average productivity of labour increases but the wage rate decreases	Rejected	Undecided (no distinct pattern)	Accepted
H_4: Labour is the "soft input" for acquiring more profit	Undecided (no distinct pattern)	Rejected	Accepted

Thus, the data analysed separately for the three regions of India give three distinct pictures of relationship between farm profit and labour exploitation. In the case of Punjab, large farmers acquire relatively more profit by efficiently utilising the different inputs so that the productivity of land is substantially higher than the small farmers. Surplus appropriation by the large farmers in Punjab is not guided by the criterion of labour exploitation, but by criterion of maximising output through intensifying the different factor costs of production.

In the case of Assam, however, the picture is different. Here, larger holdings do not necessarily maximise profit mainly for the reason that they fail to maximise output with the minimisation of factor costs. On the contrary, medium sized holdings maximise both output and profit not with the help of efficient utilization of farm inputs but mainly with the help of greater amount of labour exploitation.

The West Bengal situation falls in between these two. Here no distinct pattern of relationship can be observed in respect of making higher output and profit. However, it is clear from the data that compared to the various agricultural inputs, labour input is utilised by the big farmers of West Bengal with relatively less intensity. But payment to the labourers made by the big farmers in any case are not lower than the payments made by the small farmers. Thus, the phenomenon of labour exploitation for appropriating surplus value is not pervasive in the case of West Bengal.

From the above observations we may now conclude that the marginalist model which is used to explain the phenomenon that as farm size increases, labour gets employed with increasing quantities of land and capital, the net result would be diminishing yield and increasing profit per hectare on one hand, and diminishing output per labour but increasing labour exploitation on the other does not seem to work unrestrictedly over various regions in India.

Part II
West Bengal Agriculture

6

Performance of West Bengal Agriculture

1. Introduction

By the 1960s, West Bengal was caught up in crises of political instability. Industrial strikes, food movement, land-grab agitations etc., paralyzed normal economic and political life of West Bengal in the 1960s. In the late 1960s, two major events occurred in the political scene of West Bengal: (i) Marxist revolutionary groups scored striking successes in organizing extraparliamentary 'mass struggles' towards fighting against economic exploitation and oppression; and (ii) The Congress Party of West Bengal failed for the first time since Independence to gain a majority in the State Legislative Assembly, a United Leftist Election Committee led by the CPI(M) showed striking results. Although CP1(M) led Leftist Parties came into the state power in 1967, they got little time to continue. As a matter of fact, before 1977, although the CPI(M) led Leftist Front was considered the major political force in West Bengal, they did not get a scope to continue in the state power for a very long time. Before 1977, the political atmosphere in West Bengal was very much unstable. However, after 1977, the stability in the political atmosphere of West Bengal has come and the CPI(M) led Leftist Front has continued in the state power since 1977.

Thus, the last 10 to 15 years has been a major transitional phase in West Bengal politics. In this context, it will be interesting to examine as to how the West Bengal economy has moved during the period of political instability and what has been the picture of West Bengal economy after 1976.

To answer these queries, we need comprehensive data regarding output, earning, employment etc., with respect to each sector viz., agriculture, industry etc., of the West Bengal economy. Due to the non-availability of data of all the sectors in details other than agriculture, this chapter attempts to study the growth of agriculture in West Bengal during the period from 1960-61 to 1974-75 in particular and from 1960-61 to 1980-81 in general.

2. Growth Trends

Several attempts have been made by the economists of our country to study the growth trends of Indian agriculture [Mitra 1968; Minhas and Srinivasan 1968; Rudra 1970; Dey 1975; Sen 1982; Sau 1983 etc.]. The authors used different methods of analysis and got dissimilar results. We seek to open the question in the case of West Bengal agriculture.

The Bureau of Applied Economics and Statistics, Government of West Bengal, provides data on Index Numbers about the area, production and land productivity of foodgrains in West Bengal. In our analysis, the index number series (base period 1956-57) for the 15 years (1960-61 to 1974-75) have been used. The data are presented in Table 1.

From the data presented in Table 1, we have computed point growth rates, defined as

$$\frac{Yt+1-Yt}{Yt} \text{ X } 100$$

The series of the year to year growth rates have been presented in Table 2. Also, the average annual compound growth rates of area, output and productivity of foodgrains have been computed and presented in the same table. Annual compound growth rates have been calculated on exactly the same line as those carried out by Sau in his 1983 paper. Thus we have fitted a regression equation of the type, $\log y = a + bt$, with the time-series data of Table 1. Estimates of trend growth rate (per cent) are found from coefficient b times 100.

TABLE 1

Index Numbers of Area, Production and Land Productivity of Foodgrains (Base : crop year 1956-57=100)

Year	Area under foodgrains	Production of foodgrains	Yield of foodgrains per hectare
(1)	(2)	(3)	(4)
1960-61	105.8	118.4	111.9
1961-62	102.7	104.6	101.9
1962-63	102.4	96.8	94.5
1963-64	104.6	117.3	112.1
1964-65	107.3	124.8	116.3
1965-66	106.5	103.2	101.6
1966-67	107.7	106.3	98.7
1967-68	107.6	113.5	105.5
1968-69	111.3	128.5	195.5
1969-70	115.4	136.5	118.3
1970-71	116.7	143.2	123.4
1971-72	118.0	150.5	127.5
1972-73	117.0	130.4	111.5
1973-74	121.3	133.0	109.6
1974-75	127.3	151.5	119.0

Source: *Statistical Abstract, West Bengal,* New series no. 1

TABLE 2

Point Growth Rates and Average Annual Compound Growth rate

Year	Area under foodgrains	Production of foodgrains	Yield of foodgrains per hectare
(1)	(2)	(3)	(4)
1960-61	—	—	—
1961-62	—2.93	—11.66	– 8.94
1962-63	—0.29	—7.46	—7.26
1963-64	2.15	21.18	18.62
1964-65	2.58	6.39	3.75
1965-66	—0.75	—13.30	—12.64
1966-67	1.13	—1.76	—2.85
1967-68	—0.09	6.77	6.89
1968-69	3.44	13.22	9.48
1969-70	3.68	6.23	2.42
1470-71	1.13	4.91	4.31
1971-72	1.11	5.10	3.32
1972-73	—0.85	—13.36	—12.55
1973-74	3.68	1.99	—1.70
1974-75	4.95	13.91	8.58
Average annual compound growth rate	0.60	1.03	0.44
Trend equations : (log y=a+bt)	a= 2.0462 b= 0.0060 r= 0.94	a= 2.0887 b= 0.0103 r= 0.80	a= 2.0444 b= 0.0044 r= 0.53

The following observations can be drawn on the basis of our results of analysis presented in Table 2.

(i) The linear trend gives a growth rate of 0.60 per cent per annum in the case of area under foodgrains. 1.03 percent per annum in the case of production of foodgrains, and 0.44 per cent per annum in the case of yield of foodgrains per unit of land. These rates appear to be too much below that of all India figures worked out by different scholars.

(ii) Year-to-year growth rates show very wide fluctuations for all the items under study. However, an interesting thing that may be noted from the series is that the growth rates under each item are positively high during the period from 1968-69 to 1971-72, in which major political disturbances occurred in West Bengal.

Now, in order to test whether the growth rate is steady or rising or falling, we regress growth rate 'g' on time 't' and consider the sign of the regression coefficient of 't'. These results are presented iu Table 3.

TABLE 3

Trend of Growth Rates : Equation, g=g a+ bt (computed from Table 2)

Dependent variable : growth rate of	Regression Coefficient a	b	Correlation coefficient r
(1)	(2)	(3)	(4)
Area under foodgrains	1.2102	0.2843	0.54
Production of foodgrains	2.0365	0.5229	0.21
Yield of foodgrains per hectare	0.6958	0.2384	0.11

Equations for the trend of growth rates, presented in Table 3, do not provide a clear picture regarding the growth process in the agricultural economy of West Bengal. This indeterminancy arises because of wide fluctuations of point growth rates—the coefficient of Variations being 157 per cent in the case of area under foodgrains, 446 per cent in the case of production of foodgrains, and 1059 per cent in the case of yield of foodgrains per unit of land.

These finding suggest that nothing definite can be said about the trends of growth in West Bengal agriculture.

3. Performance of West Bengal Agriculture

An interesting finding that has come out from our analysis is that the average annual growth rate either of foodgrains production or of yield per unit of land seems to be very much lower compared to the population growth (above 2 per cent per annum). This finding indicates that the performance of West Bengal agriculture particularly for the period 1960-61 to 1974-75 is far from satisfactory from the point of view of meeting minimum food needs of the rising population. The situation remained more or less the same for the period 1975-76 to 1980-81, although the West Bengal economy overcame the political instability in this period (Table 4). Thus, Table 4 shows that the per capita availability of foodgrains per annum has not at least increased over the period 1975-76 to 1980-81 in comparison to the previous years.

TABLE 4

Availability of Per Capita Food Grains in West Bengal 1960-61 to 1980-81

Year	Total population (in '000 number)	Total food grains production (in '000 tonnes)	Annual per capita availability of food grains (in tonnes)
(1)	(2)	(3)	(4)
1960-61	34,926.3	5,936.0	0.17
1961-62	35,779.5	5,225.1	0.15
1962-63	36,652.5	4,868.7	0.13
1963-64	37,546.9	5,827.0	0.16
1964-65	38,463.0	6,259.1	0.16
1965-66	39,401.5	5,448.3	0.14
1966-67	40,362.9	5,377.2	0.13
1967-68	41,347.8	5,741.0	0.14

1	2	3	4
1968-69	42,356.6	6,597.4	0.16
1969-70	43,390.1	7,011.3	0.16
1970-71	44,312.0	7,493.7	0.17
1971-72	45,238.1	7,855.9	0.17
1972-73	46,183.6	6,772.3	0.15
1973-74	47,148.8	6,884.9	0.15
1974-75	48,134.3	7,886.0	0.16
1975-76	49,140.3	8,591.6	0.17
1976-77	50,167.3	7,493.8	0.15
1977-78	51,215.8	8,970.2	0.18
1978-79	52,286.2	8,043.4	0.15
1979-80	53,379.0	7,062.0	0.13
1980-81	54,494.6	8,281.4	0.15

Source of data : (i) *Statistical Abstract, West Bengal*, New Series : No. 1, Govt. of West Bengal.

(ii) *Estimates of Area & Production of Principal Crops in West Bengal, 1979-80*, Directorate of Agriculture, West Bengal 1981.

Note : For obtaining population estimates of each year, we have used the method of interpolation by taking the actual 1961, 1971 and 1981 Census figures of total population in West Bengal.

To examine this phenomenon a little more, an analysis of data on area and yield of foodgrains in terms of absolute numbers for the three Census years has been presented (Table 5).

It is seen from Table 5 that the area under foodgrains rose at the rate of 59.29 thousand hectares per year in the sixties, but negatively declined in the seventies. Total yield of food-grains on the other hand, increased at the rate of 155.78 thou

TABLE 5

Change in Acreage and Output in the Sixties and in the Seventies

Year	Area under foodgrains ('000 hect.)	Total yield of foodgrains ('000 tonnes)	Decennial change in Area ('000 hect.)	Decennial change in yield ('000 tonnes)
(1)	(2)	(3)	(4)	(5)
1960-61	5,530.9	5,936.0 (1.07)	—	—
1970 71	6,123.8	7,493.7 (1.22)	+592.9	+1,557.7
1980 81	6,099.3	8,281.4 (1.36)	—24.5	+787.7

Note : Figures in parenthesis indicate yield per hectare

Source : *Ibid*

and tonnes in the sixties and 78.77 thousand tonnes in the seventies. Although absolute acreage under foodgrains production has decreased in the seventies; absolute yield of foodgrains has in fact not decreased. Relatively higher productivity of land has contributed to increase the total output of foodgrains in the seventies. Modern agricultural technology (e.g., HYV seeds, fertilizers, irrigation etc.,) has possibly helped to increase the yield per unit of land in the seventies.

Although higher productivity of land has increased the absolute output of foodgrains in the seventies, the change in output relative to the previous decade has been less owing to the significant decrease in area under cultivation. Let us now see for which of the years in the seventies, this change was more acute or vice-versa.

For this purpose, we have arranged data into the interval of five-years' average beginning the year 1960-61. The data for the year 1980-81 have been taken into a separate account. The results of our analysis are presented in Table 6.

TABLE 6

Area and Production of Foodgrains : 1960-61 to 1980-81 (computed on the basis of five-years' average)

Year	Area under foodgrains (in '000 hect.)	Total yield of food-grains (in '000 tonnes)	Change in (relative to the previous group of Years) Area	yield
(1)	(2)	(3)	(4)	(5)
1960-65	5463.6	5629.2 (1.03)	—	—
1965-70	5745.2	6035.0 (1.05)	+281.6	+405.8
1970-75	6275.2	7374.6 (1.18)	+ 530.0	+1339.6
1975-80	6383.6	8032.2 (1.26)	+108.4	+657.6
1080-81	6099.3	8281.4 (1.36)	—284.3	+249.2

Note : Figures in parcnthesis indicate yield per hectare
Source : *Ibid*

Table 6 shows that both the cultivated area and yield of foodgrains increased considerably over the period 1970-75 compared to the previous years. It is also revealed from the data that both the area and production of foodgrains tended to decline after 1975-76 relative to the previous years. The decline was more drastic in 1980-81 than the previous years, and within the cultivated area in particular.

This finding clearly suggests that the so-called political disturbances in the country-side of West Bengal in the late 1960s and in the early 1970s could not affect adversely the growth of agriculture in West Bengal. What is really found from the data is that the growth of agriculture in West Bengal is moving towards diminishing order particularly after 1975-76 since when the stability in the political atmosphere in the this is claimed.

REFERENCES

Dey. Amal Krishna (1975) "Rates of Growth of Agriculture and Industry", *Economic and Political Weekly*, Review of Agriculture, June 24-28.

Minhas, B.S. and Srinivasan, T.N. (1968) "Food Production Trends and Buffer Stock Policy", *The Statesman*, November 14.

Mitra, Ashok (1968) "Bumper Harvest Has Created some Dangerous Illusions", *The Statesman*, October 14-15.

Rudra, Ashok (1970) "The Rate of Growth of Indian Economy", in Kandy Conference Proceeding, edited by E.A.G. Robinson and M. Kidron, Mac Millan, London.

Sau, Ranjit (1983) "Growth, Stagnation and Fluctuation in Indian Economy", *Economic and Political Weekly*, February 12.

Sen, Amartya (1982) "How Is India Doing"? *The New York Review of Books*, Christmas issue, December 16.

7

Types of Tenure and Farm Efficiency in West Bengal

1. Theoretical Background

There are two distinct schools of thought in the expression of "Farm Efficiency", a Marginalist and a Marxist thought[1]. The Marginalist school explains the issue of relative efficiency of different types of tenure in the following way: (a) productive efficiency is at variance as between the three types of farms namely, capitalist farms cultivated mainly by the hired labourers, peasant farms (owner cultivator) cultivated mainly by family labourers, and tenanted farms cultivated on crop-sharing basis; (b) owner cultivators are more efficient than capitalist farmers in the sense that the former put in more of labour per hectare and extract a higher output from land than the latter; (c) on the basis of same criterion as well as institutional barriers involved in the tenant cultivation the sharecropping farms are less efficient than the owner cultivating farms.

Marxist scholars, however, employ different analytical framework to examine the resource-use efficiency of farmers. According to this farmework relative efficiency is measured between large-seale and small-scale farms with respect to their input- output relations which can explain efficiently the phenomenon of differentiation or integration among the peasantry. Thus, if the law of increasing returns to scale operates in agriculture, the large-scale farms become superior to small-scale ones and as a consequence, there is a continuous differentiation among the peasantry. the reverse is true when the law of decreasing returns to scale operates.

With the above theoretical considerations in mind, an attempt has been made in this chapter to examine some inter-relations between different kinds of input and output for a composite understanding of the problem of 'farm efficiency' in agriculture.

2. Data and Methodology

We have used data of "Study on Farm Management and Cost of Production of Crops: West Bengal", published by the Socio-Economic and Evaluation Branch of the Department of Agriculture and Community Development, Government of West Bengal (henceforth we will refer FMS-CPC data).

FMS-CPC data are collected for every year beginning 1962 from some districts in West Bengal namely, Birbhum, Burdwan, Hooghly, 24-Parganas, Nadia etc. In the first year of survey (1962), there were 168 farms under the study in 21 villages. The number of farms came to 240 in 30 villages in the year 1963-64. Afterwards, during 1972-73 and 1973-74 the number of farms had increased to 576 in 72 villages spread over seven districts viz. 24-Parganas (North), Nadia, Burdwan, Birbhum, Murshidabad, Hooghly and Darjeeling hill areas.

The study covered the irrigated zones on the west of the river Bhagirathi and on the east of the Bhagirathi river. Besides these, the Darjeeling district was selected in 1973-74 with a view to assessing the pattern of cultivation and to study the economic position of cultivators of hilly areas.

The villages of each district described above were selected on the basis of their being representative of the areas in which they were situated. In selecting villages such characteristics as size of the village, kinds of crops grown, willingness of the village people to co-operate were taken into consideration. The villages selected were all accessible throughout the year.

Data were collected upon the following main items: (a) crops grown in acre, (b) land holding (ownership as well as operational), (c) family labour supply, (d) value of crops harvested, (e) cost of cultivation by human labour (family & hired), bullock, seeds, fertilizers, manures, feeds, market charges, irrigation rates, rent, depreciation, repair charges etc.,(f) farmer's earning, etc.

We have used in this study the FMS-CPC data pertaining to the years 1965-66 and 1973-74 for the five districts of West Bengal viz. Birbhum, Burdwan, Hooghly, 24-Parganas and Nadia. We have selected two different time-points with a view to understanding the effect of so-called new technology which has been started to introduce in agriculture in 1965, but as per some social scientists it has subsequently spreaded over various regions in West Bengal in the 1970s. For this specific reason, the districts which were newly introduced by the FMS-CPC authority in the 1970s, have been excluded from our present exercise. The distribution of selected farms by each district at two time-points is given below on the basis of which our present exercise is carried out.

Table 1

Distribution of Farms by District

District	Total no. of selected farms in the year 1965-66	1973-74
(1)	(2)	(3)
Birbhum	24	96
Burdwan	32	96
Hooghly	16	96
24-Parganas	24	95
Nadia	24	64
Total	120	447

It may be noted that we have used holding-wise disaggregated data for the farms presented above. The holding wise data are available in the Part I report of FMS-CPC for every year. Our results in this exercise are based on those farms for which disaggregated data are available. We now propose to present our results of analysis within the framework discussed earlier.

3. Empirical Validity of Marginalist View

We shall first examine the marginalist view with regard to farm efficiency on the basis of data discussed above. For this purpose, we have classified the selected farms into two broad categories: (i) Owner cultivator (having land over which he has sole ownership rights and cultivation done with the help of both hired and family labourers), (ii) Tenant cultivator (mainly no ownership rights over the land which he cuitivates and cultivation done mainly on the basis of family labour).

The distributions of owner and tenant cultivating farms with respect to each selected district are given in Table 2.

TABLE 2

Distribution of Farms by Types of Tenure

District	Owner cultivating farms		Tenant cultivating farms		Total farms	
	1965-66	1973-74	1965-66	1973-74	1965-66	1973-74
(1)	(2)	(3)	(4)	(5)	(6)	(7)
Birbhum	14	77	10	19	24	96
Burdwan	5	73	27	23	32	96
Hooghly	9	80	7	16	16	96
24-Parganas	16	75	8	20	24	95
Nadia	4	22	20	42	24	64
Total	48	327	72	120	120	447

Table 2 shows that in 1965-66, out of 120 selected farms 72 were tenant operated whereas in 1973-74 out of 447 selected farms, only 120 farms were tenant operated. It seems to us that as per the selected holdings, very little tenanted farms (in terms of proportion) has been caught in the sample net in 1973-74 compared to the year 1965-66. Considering all the limitations of comparisons between these two sets of data, we nevertheless believe that less incidence of tenant operated farms in the 1973-74 selection would not be just as selection causalty. In fact, FMS-CPC selected holdings stratifying by size-classes and not by the tenurial classes. It will be therefore misleading to think that decrease in the proportion of tenant cultivated holdings in the sample units in 1973-74 has been possible due to selection error. There is certainly a fall in tenant cultivation in the districts of West Bengal and this has occurred due to the so-called new technology. Extensive development of irrigation (including Govt. canals) as well as subsidised supply of inputs like chemical fertilizers, HYV seeds etc., have encouraged the cultivators to cultivate their land themselves. Thus, the virtual elimination of a particular mode (tenant cultivation) and its replacement by another mode (owner cultivation with hired labourers) that has been observed by some other scholars also may be accepted as a general trend of West Bengal agriculture.

Within this kind of trend in West Bengal agriculture, let us examine the relative efficiency of the two modes of cultivation—owner cultivation and tenant cultivation—in terms of input-output relationship at two different points in time (Table 3).

If we assume that putting in more of labour and more of non-labour material inputs on land and extracting higher output from per unit land are the indicators of farm efficiency, the data examined in Table 3 do not indicate significant differences between owner and tenant cultivations for the years 1965-66 and 1973-74. In fact, the differences observed in the data on three items viz. output, labour input and material input per acre between owner and tenant cultivated farms during 1965-66 to 1973-74 are so small that one cannot make a definite conclusion regarding the farm efficiency between the two modes of cultivation. However, judging by the each district it follows from the same table that except Nadia in 1965-66 and Burdwan in 1973-74, there are no significant differences between owner and tenant cultivated farms in terms of input-output variations for the other districts under study. In Burdwan, owner cultivators seem to be more efficient than tenant cultivators in 1973-74 but not in 1965-66. Conversely, in Nadia district, owner cultivators seem to be efficient in 1965-66 but not in 1973-74. Both the districts in the recent years are regarded as so-called developed areas in terms of irrigation, cropping pattern etc, but

TABLE 3

Output material cost and labour cost per acre by types of tenure in some districts of West Bengal between 1965-66 and 1973-74

Name of the district	Output per acre (in Rs.) 1965-66	1973-74	Material cost per acre (Rs.) 1965-66	1973-74	Labour cost family 1965-66	1973-74	per acre (Rs.) hired 1965-66	1973-74
1	2	3	4	5	6	7	7	9
				Owner cultivator				
Birbhum	650.24	1239.56	229.10	360.42	50.11	85.80	150-23	212.91
Burdwan	850.40	2093.74	208.03	685.44	90.31	121.40	171.88	443.81
Hooghly	1261.90	2827.95	337.50	594.40	117.25	148.52	165.76	361.19
24—Parganas	1313.21	1965.07	232.30	333.82	135.21	176.15	209.96	264.05
Nadia	5555.05	1730.65	98.54	338.72	76.30	104.28	76.00	297.01
All districts	936.64	1978.28	229.19	485.47	91.00	126.97	162.11	316.97
				Sharecropper				
Birbhum	557.49	1104.48	143.58	291.07	40.31	100.01	85.21	148.86
Burdwan	909.28	1624.95	228.84	424.64	63.06	164.10	118.84	261.28
Hooghly	1636.27	2888.26	425.01	659.90	113.91	244.09	239.62	429.20
24—Parganas	1423.28	2055.09	256.51	397.50	149.00	192.66	193.71	291.44
Nadia	732.67	1795.10	126.34	316.52	90.36	177.38	106.31	222.15
All districts	911.65	1820.95	208.87	391.12	77.45	171.81	126.97	254.21

Note : Material input includes seeds, manure, fertilizers and irrigation.

opposite result betwern these two districts in 1973-74 seem to be somewhat puzzling. Thus, a general remark about the efficiency or inefficiency of two types of tenure cannot possibly be drawn on the basis of the incidence of modern inputs spreading over various types of farms in different regions.

We may now turn to discuss the marginalist model which is used to explain the phenomenon that as farm size increases, labour gets employed with increasing quantities of land and capital, the net result would be diminishing yield per hectare and diminishing output per labour.

Diagram-1: Output per acre against farm size by types of tenure. West Bengal 1965-66 & 1973-74

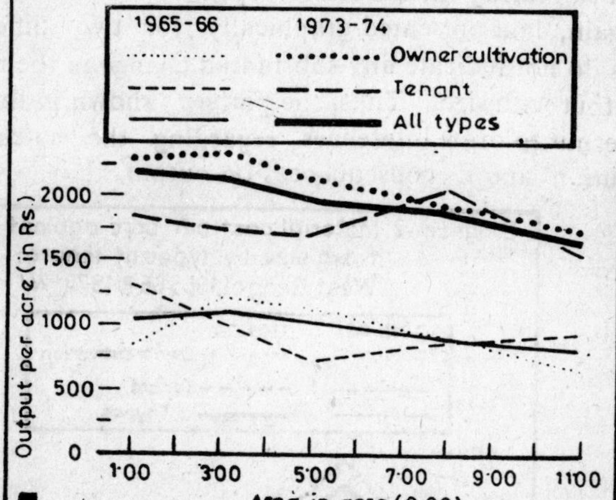

We shall be presenting a certain number of graphs representing the pattern of dependence of yield per acre in value terms of all crops taken together by types of tenure with the primary purpose of illustrating certain arguments that are put forward by Marginalist theeorists. The data graphically presented refer to pooled data of five districts of West Bengal at two different time-points.

Attention may first be drawn to the Diagram 1. It is seen that one cannot discover any tendency of output per acre decreasing or increasing with farm size either in the owner operated farms or in the tenant operated farms for both the points of time. The same, however, cannot be discerned in the case of combined category for both the time points. It is seen that output per acre increases, although not very rapidly, in the combined category for the both the points of time.

It seems to us that the inverse relation between size and productivity does not necessarily mean that the relationship holds in all ranges of the size variable for each and every type of tenure[2]. In fact, our data show that when it comes to individual tenure, there are no discernible systematic patterns between output and size. This pattern, however, does not hold true when all types of tenure are combined into one group and accordingly data are pooled in it against each farm size group. It is thus clear that such pooling or aggregation can falsify the nature of dependence of farm size.

Again, data presented graphically for two different time-points do not indicate any substantial change in the movements of output with size. Thus, the pattern shown in the graph do not permit to draw inferences regarding the so-called 'green revolution' and its consequences on output.

Diagram-2: Material cost per acre against farm size by types of tenure: West Bengal, 1965-66 & 1973-74

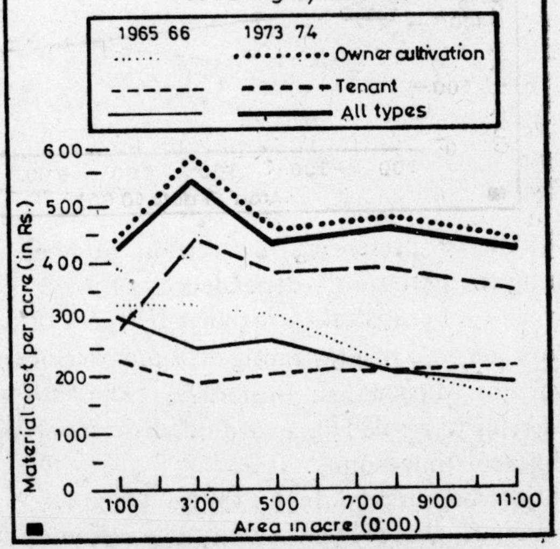

Coming now to the results of our analysis presented in Diagrams 2 & 3 relating to the intensities of labour and material inputs with corresponding farm size-groups under different types of tenure for the years 1965-66 and 1973-74, the following points may be noted.

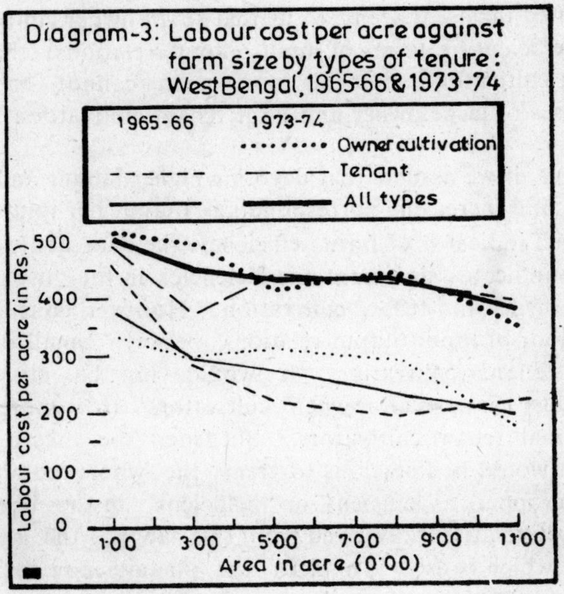

(a) There is a tendency of decreasing material input per unit of land with farm size increasing for farms either based on owner cultivation or on tenant cultivation during 1965-74. The combined category also gives similar result.

(b) Labour cost per acre decreases with farm size in the farms of combined category. But the picture is not so clear in the cases of individual tenurial type. It may be noted in this context that we have not been able to present labour input data in terms of mandays or manhours due to lack of information. However, we feel that intensity of labour input measured in terms of mandays or manhours may give better results than the procedure adopted in the FMS-CPC study.

(c) There is a difference in the utilization of labour as well as material inputs per acre among the farmers of corresponding farm size groups, more particularly in the smaller size-groups, under owner and tenant cultivated farms for both the points of survey. There is also a difference in the productivity of land in the smaller size groups between owner and tenant cultivated farms. It seems to us that small owner cultivators are more efficient (in terms of input-output relations) than small tenant cultivatoıs. But this conclusion cannot be drawn in the cases of large owner and large tenant cultivators.

Thus, if we assume that devoting higher labour and material inputs and increasing corresponding output per unit of land is the chief indicator of farm efficiency, the data analysed so far do not indicate signifcant inefficiencies in the use of land under owner and tenant cultivation. However, considering the behaviour of input-output relations between small owner and small tenant cultivatiors, it would not be misleading to conclude that small owner cultivators are more efficient than small tenant cultivators. But from the macro point of view it would be fallacious to treat the whole community of share-croppers as 'efficient' or 'inefficient' in the use of land under cultivation.[3] Viewed from this angle, the marginalist model which is used to explain the phenomenon that as farm size increases labour gets employed with increasing quantities of land and capital the net result would be diminishing yield per acre and diminishing output per acre and diminishing output per labour does not seem to work unrestrictedly over various types of tenure and size groups.

4. Empirical validity of Marxist Approach

We may now put together our findings with regard to the different interrelations for examining the Marxist approach on farm efficiency.

TABLE 4

Input-output ratio by types of tenure in West Bengal between 1965-66 & 1973-74

Size groups	Input-output ratio by types of tenure Owner		Tenant		All types	
	1965-66	1973-74	1965-66	1973-74	1965-66	1973-74
(1)	(2)	(3)	(4)	(5)	(6)	(7)
Below 2	0.769	0.415	0.557	0.425	0.676	0.416
2—4	0.515	0.501	0.465	0.529	0.484	0.507
4—6	0.529	0.427	0.554	0.469	0.537	0.439
6—10	0.562	0.466	0.440	0.410	0.477	0.451
Above10	0.484	0.487	0.465	0.426	0.471	0.477
All	0.515	0.470	0.453	0.449	0.477	0.465

Note: Total input includes both labour and material input (in Rs.)

Results of our analysis presented in Table 4 suggest that:

(a) there is no particular tendency to the existence or non-existence of increasing returns to scale for inputs actually used in agriculture;

(b) the various tenure farms can have different effects on production.

At this point in the study, West Bengal data show that the relevance of machinery is limited for the less developed districts described above with cheap labour, returns to scale are likely to be constant until capital becomes cheaper relative to labour. If returns to scale are constant the small farms may exist without loss of efficiency. But our previous analysis shows that the extent of share cropping declines over time and the ranks of agricultural labourers get swelled. Thus we find that the Marxist formulations regarding farm efficiency and its consequences on labour and tenancy markets do not seem to work unequivocally in the case of West Bengal.

NOTES

1. In this context see the discussion the results derived by N.K. Chanda (1974).

2. See Chattopadhyay and Rudra (1976) for an extensive discussion of the point.

3. This agrees with the conclusion drown by the present author in 1979 (Chattopadhyay 1979).

REFERENCES

Chandra, N.K. (1974): "Farm Efficiency under Semi-Feudalism. A Critique of Marginalist Theories and Some Marxist Formulations", *Economic and Political Weekly*, Special Number, August.

Chattopadhyay, Manabendu and Rudra, Ashok (1976): "Size-Productivity Revisited", *Economic and Political Weekly*, Review of Agriculture, September 25.

Chattopadhyay, Manabendu(1979): "Relative Efficiency of Owner and Tenant Cultivation: A Case Study", *Economic and Political Weekly*, Review of Agriculture, September 29.

8

Structural Constraints of Rural Development

1. Concept

Rural development would essentially mean desired *positive change* in the rural areas both in a quantitative as well as qualitative sense. It integrates a variety of elements of human life and activities. We may broadly classify the numerous elements into the following broad dimensions : (a) Social, (b) Economic, (c) Technological and (d) Natural. Rural development means desired changes in all these elements. But such changes should take place in a mutually supporting relationship so as to generate an optimum development. If certain aspects of life such as economic and technological are developed, but other aspects especially social and natural remain backward, then such a development may be called partial development, not of course optimum development. From this point of view, rural development means development (quantitative and qualitative change) of the rural areas in such a way that each component of rural life changes in a desired direction and in sympathy with the other components (For an extensive discussion of this point, see Misra and Sundaram 1979).

Although the concept of rural development covers a variety of activities of human life, the central concept of rural development presented here is of a process through which rural poverty is alleviated by sustained increases in the productivity and income of low-income rural workers and households. The other aspects of rural development have not been possible to incorporate here due to lack of data in a wide-scale.

Operationally, this concept of rural development requires that target groups be specified among the rural households, for whom specific measures to raise production and income can be designed, and in those cases the resulting flow of benefits-direct and indirect are both identifiable and potentially measurable (In this context, see the paper, World Bank 1975). However, one approach for identifying target groups is based on the identification of agrarian class categories in terms of ownership of the relevant factors of production such as land, labour and capital.

2. Agrarian Structure

Agricultural households of rural India can be classified into three major categories according to the ownership of land : (i) Owners (possessors solely of own land), (ii) Tenants (possessors mainly of land belonging to others), and (iii) Agricultural labourers (possess no cultivated land either owned or leased in, receive wages for work in agricultural production).

The National Sample Survey (NSS) has been virtually an important source of analysing land-holding data over various time points at all-India level. It provides data that ralate to the proportion of households and area operated by size of operational holdings.

Data on ownership of land resources during the 1950s reveal that about 47 per cent of the population owned either no land or less than one acre of land and accounted for about 1.38 per cent of the total land resources as can be seen from Table 1.

Since the structure of land ownership had failed to meet the ends of social justice, it was felt that the best course in bringing reduction in inequalities of income was to bring about a reduction in inequalities of land ownership. To this end, ceilings were imposed on larger holdings and the surplus land

TABLE 1

Percentage distribution and estimated no. of households and area owned by size-class of household ownership holding (1954-1955)

Size-class of household ownership holding (acres)	No. of households ('000)	Area owned ('000 acres)	Percentage of households	Percentage area owned
(1)	(2)	(3)	(4)	(5)
*0.00	14444	—	22.00	—
0.01-0.99	16346	4275	24.89	1.38
1.00-9.99	26536	107231	40.42	34,61
10.00-39.99	7451	135501	11.35	43.73
40.00-99.99	764	42218	1.16	13.63
100.00-499.99	114	18512	0.17	5.97
500.00 & above	4	2113	0.01	0.68
Total	65659	309850	100.00	100.00

*Households owning no land or less than 0.005 acres are shown against this size-level

Source : National Sample Survey 8th round; July 1954—March 1955—*First Report on land holdings, Rural Sector*

was sought to be distributed among the small and marginal peasants. The rational of this being that such peasants would then be able to meet their consumption requirements from this augmented size of holding and thus would be able to move above the poverty level.

This hope seems to have been frustrated as can be seen from the existing data on land distribution (Table 2). The degree of concentration as measured by the Lorenz ratio is estimated to be 0.71.

TABLE 2

Percentage distribution and estimated no, of households and area owned by size-class of household ownership holding (1971-72)

Size-class of household ownership holding (acres)	No. of households ('000)	Area owned ('000 acres)	Percentage of households	Percentage of area owned
(1)	(2)	(3)	(4)	(5)
*0.00	7558	—	9.64	—
0.01—0.99	27609	6106	35.23	2.07
1.00—9.99	35405	130853	45.18	44.28
10.00—24.99	6138	90846	7.83	30.74
25.00—49.99	1342	44460	1.71	15.05
50.00 & above	318	23235	0.41	7.86
All	78370	295500	100.00	100.00

*Households owning no land or less than 0.01 acres are shown against this size-group

Source : National Sample Survey : 26th round, July 1971—September 1972. *Tables on land holding—All India*

This high value of the ratio implies that the heavy concentration of land ownership has remained practically unaltered in spite of the various legislative measures adopted, particularly in the 1950s. On the other hand, the emerging structure has perpetuated the inequitable distribution of economic resources.

The agrarian structure as it stands today reveals that on the one hand the inequality in the distribution of land ownership has not substantially altered in favour of the rural poor through land reforms, on the other hand qualitative changes have taken place in the mode of operation of land as a result of commercialisation of agriculture and penetration of new technology into the countryside which in many cases have worsened the position of the rural poor in the agrarian structure (Hanumantha Rao 1975). The new technology, being market oriented and capital intensive has mainly favoured the big peasants; the small farmers not possessing the required resource base or requisite knowledge and being a risk avertor have not by and large, opted for this technology (Dasgupta 1980). Small and marginal farmers are therefore placed at a disadvantage vis-a-vis the big owners and so a few options remain open to them; (a) to cultivate their land in the traditional way so as to provide a meagre subsistence, (b) to sell or lease out their land to big owners and seek employment elsewhere in the absence of which they are forced to join the rank of landless labourers.

Let us now look at the changes in the extent and pattern of tenant cultivation in India.

Available evidence suggests that the area under tenancy declined absolutely and also in relative terms, from 20.34 per cent of the total operated in 1953-54 to 10.57 per cent in 1971-72. Most of this, however, had been achieved before 1960, as may be assessed from Table 3.

Comparing between 1960-61 and 1970-71 (17th and 20th roundsof the NSS), the following observations can be made regarding tenants and labourers engaged in agriculture.

(a) The extent of tenancy seems to have remained the same on an all-India level during 1960-61 to 1970-71

(b) Percentage of tenancy holdings cultivated by the landless tenants has decreased significantly between 1960-61 and 1970-71, while there is a large increase (51 per cent) in the number of agricultural labourers during this period. It is interesting to note that this increase has been possible due to demographic factor by 22 per cent only. Hence, this evidence strengthens clearly the possibility of increasing number of agricultural labourers due to the larges cale eviction of landless tenants.

TABLE 3

Comparative Tenancy and Agricultural Labourers in 1960- and 1970-71 : All India

Items	1953-54	1960-61	1970-71
(1)		(2)	(3)
i) Percentage of operated area leased in	20.34	10.70	10.57
ii) Percentage of holdings reporting some leased in area cultivated by landless tenants	—	21.85	15.03
iii) No. of agricultural labourers (in lakhs)	—	314.8	474.9
iv) Relative increase of agricultural labourers	—		50.9
v) Increase due to demographic factor			22.0

From these evidences we may conclude that tenancy declined very rapidly in the 1950s, but in the 1960s the intensity of decline is not substantial. Hence the decline in tenancy between 1953-54 and 1971-73 may not entirely be attributed to the land reform measures. Rather the shift towards owner cultivation from tenant cultivation occurred in the 1950s due to land reform measures, but in the 1960s it was due to various kinds of development measures in agriculture including the development of HYV seed technology and mechanisation. This is strengthened by the fact that a large as well as middle farms bias in the adoption of the water-seed fertilizer innovation has been noticed in the 1960s in the so-called green revolution areas. Big as well as middle farmers have decisive advantages over the smaller ones in respect of their own capital as well as credit facilities. The official policies of low incidence on agricultural incomes and assets, and of subsidisation of modernising inputs have encouraged big as well as middle landowners to take more direct interest in large-scale farming. On the other hand, economic pressures might have forced out the small farmers, especially small tenants and made them join the swelling ranks of agricultural labourers, particularly in view of the costlines and credit intensity of the new agricultural technology.

Moreover, available evidence suggests that under the impact of the intensive development strategy the landowner's income per unit of land increased between 50 and 100 per cent and that of the labour between 25 and 35 per cent [Chakravarty 1980]. Whereas in the old order of traditional techniques the relative share of the labour income in gross output was in the range of 25 to 30 per cent, in the new order the share was reduced to the range of 20 to 35 per cent [ibid]. Thus, while incomes of labourer and landowners increased in the course of agricultural development, disparity of income between these two classes also increased.

To the extent that rural masses lose their direct control over land and its operation, they also lose their capacity to fulfil their basic consumption needs on the basis of self-provisioning. This is because a small producer who was a tenant and/or an owner-cultivator was also a consumer of the product of his own small farm. Moreover, the landless labourers who were attached to the landlord on the basis of patron-client relationship and received wages in kind were also served by this framework of self-provisioning to a considerable extent. Self provisioning, which had made an important contribution to the fulfilment of basic needs in the traditional form of agriculture has come to an end with the introduction of new technology [Joshi 1982]. Thus, the customary arrangements like labourer's attachment to the landlord through patron-client relationship has been replaced by some new tying arrangements like permanent labour contracts, introduction of money wage rates etc. by the big farmers in the technologically developed areas. This has made the rural poor dependent on the market for their basic needs. Also, with the expansion of the market, the command over purchasing power became an essential condition for the fulfilment of the needs of the rural poor. But this in turn depends on regular employment opportunities and earnings through the competitive market framework. But in reality, the limited success of the employment promotion schemes has proved to be the major impediment in the upliftment of the rural masses [Ahuja 1978; Minhas 1974]. These institutional and technological changes which have taken place increased the disparities between the different categories of rural poor with the result that an accentuation of agrarian tensions and conflicts is underway in different parts of rural India.

3. Constraints

The evidences presented above show that the benefits of agricultural development (in terms of land reform measures, introduction of new technology etc.,) have been directed to a particular landowning class (i. e., big and middle farmers) and not towards all the rural classes. On the contrary, these have simultaneously created a problem of relative deprivation amongst the small peasants and labourers. This suggests that the development programmes so far taken in view of the upliftment of the rural poor are not compatible with the village power structure. The village power structure is still guided by the property structure and as a consequence the benefits of development are captured by the large landowners. Thus, the present agrarian structure which may appear to be quite rational on pure economic considerations seems to turn out as politically infeasible in such a situation where the polarised distribution of labour and capital has taken place [Chakravorty 1980]. The key to rural immiserisation, therefore, lies not noly in how the growth and stability of overall agricultural production is maintained but in the way in which resources are controlled under the existing production system to obtain profit and social power and other concomitant factors such as state policies and government measures for the upliftment of the rural poor.

REFERENCES

Ahuja, Kanta (1978) *Idle Labour in Village India*, Manohar, New Delhi.

Chakravarty, Sukhamoy *Alternatives in Agricultural Development: An ICSSR Working Group Report*, Allied publishers Pvt. Ltd, New Delhi.

Minhas, B.S. (1972) "Rural Development for Weaker Sections: Experience and Lessons", Paper presented in the Seminar on *Rural Development for Weaker Sections*, Indian Institute of Management, October (Mimeo).

Misra, R.P. & Sundram K.V. (1979) *Rural Area Development: Perspectives and Approaches*, Sterling, New Delhi.

Raj, K.N. (1976) "Trend in Rural Unemployment in India: An Analysis with Reference to Conceptual and Measurement Problems", *Economic and Political Weekly*, Special Number, August.

Rodgers, G.B. (1973) "Effects of Public Works on Rural Party", *Economic and Political Weekly*, Annual Number.

Sanyal, S.K. (1977) "Trends in Rural Unemloyment in India: Comment" *Economic and Political Weekly*, January 29.

World Bank (1975) *Rural Development*, International Bank for Reconstruction and Development, Washington (Sector Policy Paper).

Dasgupta, Biplab (1980) *New Agrarian Technology and India*. Chapter 5 Macmillan, Delhi.

Hanumanth Rao, C.H. (1975) *Technological Change and Distribution of gains in Indian Agriculture.* Macmillan, Delhi.

Joshi, P.C. (1982) "Poverty, Land Hunger and Emerging Class Conflicts in Rural India"" in S. Jones, P.C. Joshi & M. Murmis (ed); *Rural Poverty and Agrarian Reform,* Allied Publishers Pvt. Ltd.

Minhas, B.S. (1974) "Rural Poverty, Land Distribution and Development Strategy: Facts", in T. N. Srinivasan and PK Bardhan (ed.). *Poverty and Income Distribution in India.* Statistical Publishing Society, Calcutta.

Jaico